GENETIC VARIANCE DETECTION
NUTS & BOLTS OF DHPLC IN GENOMICS

Edited by
Karl Heinz Hecker, Ph.D.

DNA Press™

NEW AGE

First Indian Sub-Continent Edition : 2007
by New Age International (P) Ltd., Publishers
Original Printed in the U.S.A.

Printed in India at Mehra Offset, Delhi.
1 2 3 4 5 6 7 8 9 10
C-06-10-1090

ISBN: 81-224-1983-6
Original Edition Price: $ 32.95
Indian Edition Price: Rs. 325.00

PUBLISHING FOR ONE WORLD

NEW AGE INTERNATIONAL (P) LIMITED, PUBLISHERS
(formely Wiley Eastern Ltd.)
New Delhi · Bangalore · Chennai · Cochin · Guwahati · Hyderabad
Jalandhar · Kolkata · Lukhnow · Mumbai · Ranchi
Visit us at www. Newagepublishers.com

Contents

SECTION I: Fundamentals of DHPLC

Chapter 1
Basics of automated, high-accuracy mutation screening with the WAVE® Nucleic Acid Fragment Analysis Systems 15
Karl H. Hecker

Chapter 2
Influence of PCR on DHPLC Polymorphism Characterization 35
Michael L. Nickerson

SECTION II: Gene Specific Applications of Mutation Detection by DHPLC

Chapter 3
Denaturing High Performance Liquid Chromatography (DHPLC), an important mutation detection technique in mutation analysis of the breast and ovarian cancer predisposing genes BRCA1 and BRCA2 53
Norbert Arnold and Dieter Niederacher

Chapter 4
Mutation detection in the tuberous sclerosis genes TSC1 and TSC2 by DHPLC 79
Penelope S. Roberts, Sandra L. Dabora, David J. Kwiatkowski

Chapter 5
Rapid analysis to detect base changes in human mitochondrial DNA by restriction enzyme analysis followed by denaturing high performance liquid chromatography 93
Patricia E. McAndrew and Michael A. Marino

Chapter 6
Mutation analysis of LDLR mutations in patients with familial hypercholesterolaemia 117
Caroline F. Bunn and Peter M. George

Contents *(continued)*

SECTION III: Novel and Innovative Applications of DHPLC

Chapter 7
Novel DHPLC based strategy for identification and characterization of (bi)-clonal T-cell receptor targets for the detection of minimal residual disease in childhood acute lymphoblastic leukemia 137
Udo zur Stadt, Johannes Rischewski, Conny Eckert, Katharina Michael, Reinhard Schneppenheim, and Hartmut Kabisch

Chapter 8
SIRPH: a rapid, quantitative SNuPE-IP RP HPLC method for simultaneous methylation analysis at multiple CpG sites 151
Osman El-Maarri, Johannes Oldenburg, and Jörn Walter

Chapter 9
Isolation and sequencing of heteroduplex fragments from DHPLC analysis of somatic, low-frequency mutations 163
Jan Alsner

Chapter 10
Analysis of promoter methylation using denaturing high-performance liquid chromatography (DHPLC) 175
Beate Betz and Dieter Niederacher

Index 187

Summary of chapters

SECTION I: Fundamentals of DHPLC

Chapter 1: Basics of automated, high-accuracy mutation screening with the WAVE® Nucleic Acid Fragment Analysis Systems

The analysis of DNA fragments, PCR products or restriction fragments, by ion-pair reversed-phase high-performance liquid chromatography (IP RP HPLC) on an alkylated, nonporous poly(styrene-divinylbenzene) matrix (DNASep® Cartridge) using the WAVE® Nucleic Acid Fragment Analysis Systems is a powerful and versatile tool for DNA analysis. Resolution of DNA fragments is based on two principles, size-dependent retention of double-stranded (ds) DNA and differential retention of ds vs. single-stranded (ss) DNA. Temperature Modulated Heteroduplex Analysis (TMHA) utilizes both principles of separation to detect single nucleotide polymorphisms (SNP) and short insertions/deletions. At a given temperature the difference in the melting between homo- and heteroduplexes is revealed by differences in retention times. The temperature at which differential melting occurs is sequence dependent and is predicted accurately using either WAVEMAKER™ or WAVE Navigator™ software, which use a modified Fixman-Friere algorithm. Detection of known and unknown sequence variations can be performed on DNA fragments of up to 1,000 base pairs with high sensitivity and specificity. The use of fluorescent labels is compatible with this technology and increases sensitivity. Retention times are increased and resolution is not affected. Fluorescent labeling significantly increases sensitivity. Here we provide the reader with a conceptual overview of the principles of mutation detection by DHPLC and of the factors determining the outcome, quality, and reliability of the data generated using this technology.

Chapter 2: Influence of PCR on DHPLC Polymorphism Characterization

Mutation detection and polymorphism characterization by DHPLC requires validation of method parameters, reagents, and instrumentation. Accurate, reproducible correlation between a DNA elution profile and a polymorphism can be obtained once the elements of the experimental protocol have been carefully standardized. This is critical for downstream data processing of large numbers of chromatograms using scatter plots that sort elution profiles based on curve fitting algorithms. Extraneous influences, which affect the shape of an elution profile, may lead to incorrect sorting of chromatograms and erroneous interpretation of SNP and mutation carriers. To date, although DHPLC method design has been optimized, PCR reagents have not been examined in detail.

This is important since the large number of laboratories practicing DHPLC use a wide variety of PCR reagents. Below, DHPLC chromatograms of PCR products containing a SNP were amplified using six different polymerases and are compared. Low fidelity polymerases appear to affect DHPLC elution profiles through errors that are randomly introduced during PCR, which create extra peaks and shoulders relative to profiles obtained from DHPLC of products amplified by high fidelity polymerases. Thus, for a given SNP, elution profiles varied dramatically depending on whether a low or high fidelity polymerase was chosen for amplification. The effect of PCR annealing temperature was found to have minimal influence on signatures generated using low or high fidelity enzymes. Lastly, Random Mutagenesis-PCR is briefly presented to highlight the utility of amplicons containing errors introduced during PCR. When subcloned, these PCR products can be used to validate mutation detection procedures and DHPLC instrumentation.

SECTION II: Gene Specific Applications of Mutation Detection by DHPLC

Chapter 3: Denaturing High Performance Liquid Chromatography (DHPLC), an important mutation detection technique in mutation analysis of the breast and ovarian cancer predisposing genes BRCA1 and BRCA2

The syndrome of hereditary breast and ovarian cancer has been known for years but the identification of individuals at risk and their inclusion in surveillance programs became feasible only after cloning of both predisposing genes BRCA1 and BRCA2 in 1994/1995 and the ability to test for alterations in both BRCA genes. Testing patients at high risk in an established clinical trial requires standardized procedures in detection and interpretation of BRCA mutations and the acquisition of population-specific genetic data. In the German population a comprehensive study was initiated in 1996 by the formation of a multi-center research collaboration, the "German Consortium for Hereditary Breast and Ovarian Cancer," supported by the Deutsche Krebshilfe (German Cancer Aid). The study protocol uses a multidisciplinary approach that provides counseling and genetic testing by bringing together, in each center, a team of genetic counselors, gynecologists, psycho-oncologists and molecular biologists. In the beginning different strategies for the analysis of the BRCA genes like SSCP, PTT, and direct DNA-sequencing were used in the individual centers. However, in a unique effort the "German Consortium for Hereditary Breast and Ovarian Cancer" established DHPLC as the superior technology for the detection of genetic variance in BRCA genes and implemented the technology on ten WAVE® Nucleic Acid Fragment Analysis Systems (Transgenomic Inc., Omaha, NE) simultaneously in its centers. A critical aspect to the success of the project was the close collaboration between centers, which shared PCR primer sets, optimized DHPLC conditions, and positive controls. Establishing the technology across multiple centers, sensitivity of the technology, quality management, and results are presented and discussed in this chapter.

Chapter 4: Mutation detection in the tuberous sclerosis genes TSC1 and TSC2 by DHPLC

Tuberous sclerosis is an autosomal dominant hamartoma syndrome, which is caused by mutations in either TSC1 or TSC2. Here we review our experience with the use of denaturing high pressure liquid chromatography (DHPLC), quantitative PCR (QPCR), and long-range PCR (LRPCR) for mutation detection in a group of 660 patients with this disease. Mutations have been identified in 469 (71%) patients. DHPLC is a very sensitive and specific technique for mutation detection in our hands, with 99% mutations found by DHPLC in blinded trials, compared to 59% by conformation-sensitive gel electrophoresis and 72% by SSCP. We detected 86% of the mutations in these genes by DHPLC. The remainder was shown to be large genomic rearrangements and these mutations were detected by the other methods. Most patients with tuberous sclerosis are sporadic cases and TSC2 mutations are 4.2 times as common as TSC1 mutations. Using DHPLC, we have identified elution shifts for exon amplicons in which sequence analysis failed to identify a mutation. Cloning of these amplicons followed by analysis of individual clones is used in these cases to evaluate the possibility of mosaicism. In the last 120 patients studied, by this method we identified mosaic mutations in 4 (3%) patients. By counting the number of clones containing the mutation, we determined that the level of mosaicism in these patients is 7-26%.

Chapter 5: Rapid analysis to detect base changes in human mitochondrial DNA by restriction enzyme analysis followed by denaturing high performance liquid chromatography

Molecular analysis of mitochondrial DNA (mtDNA) has become an indispensable tool for the diagnosis of mitochondrial diseases, forensic investigations, the study of human evolution, and population genetics. However, several features of the mitochondrial genome can complicate analysis of mtDNA. We developed a method to scan the entire mitochondrial genome for base changes using denaturing high performance liquid chromatography (DHPLC). DHPLC is a rapid, versatile technology that uses an ion-pair reversed-phase column to separate heteroduplex molecules from homoduplex fragments when base changes are present. The complete mitochondrial genome was amplified with 18 overlapping sets of PCR primers. The PCR products were digested with a limited number of restriction enzymes to produce 100-600-bp fragments suitable for DHPLC analysis. A Standard Reference Material (SRM 2392) established to provide researchers with a well-characterized source of mitochondrial DNA was used to validate the feasibility and sensitivity of this approach to detect base changes in the mitochondrial genome. This method successfully detected 100% (21/21) of the single nucleotide polymorphisms distributed throughout the 16.5 kb mitochondrial genome. This report describes a rapid, sensitive DHPLC method to screen the human mitochondrial genome for base changes. This approach can be used to detect the presence of both inherited or somatic mtDNA base changes and for a variety of different applications involving mtDNA.

Chapter 6: Mutation analysis of LDLR mutations in patients with familial hypercholesterolaemia

Familial hypercholesterolaemia (FH) is a common inherited disease, with premature vascular disease occurring 10-20 years earlier than in patients with polygenic hyperlipidaemia. FH may be caused by many different mutations in the low density lipoprotein receptor (LDLR) gene and, to date, over 700 mutations have been described, most of which occur rarely and often only in single families. Although founder effects occur, countries with heterogeneous populations have an array of mutations, with the majority being point mutations; thus a gene screening approach is the appropriate first step for mutation detection. Methods previously used to screen for LDLR mutations, such as SSCP, lack sensitivity and are not easily automated in the diagnostic laboratory. We have used DHPLC for mutation prescreening, as an integral part of LDLR analysis in a clinical setting. Eight novel mutations have been identified by DHPLC from 29 different mutations (in 55 patients). Six novel mutations were localized to the EGF domain, one in the signal peptide and the eighth in the membrane-spanning domain. The majority of the 29 mutations are localized in the EGF domain and not the ligand-binding domain as previously reported. We conclude that DHPLC is the method of choice for LDLR mutation detection in a clinical setting as it can be integrated into an automated laboratory process, subsequent to DNA extraction and PCR set-up. DHPLC provides decreased sample processing time as well as increased sensitivity and specificity necessary in the clinical environment.

SECTION III: Novel and Innovative Applications of DHPLC

Chapter 7: Novel DHPLC based strategy for identification and characterization of (bi)-clonal T-cell receptor targets for the detection of minimal residual disease in childhood acute lymphoblastic leukemia

Incomplete T-cell-receptor rearrangements are widely used for detection of minimal residual disease (MRD) in childhood acute lymphoblastic leukemia

(ALL). In a substantial number of cases both alleles are rearranged and PCR products have either to be cloned or excised and re-amplified from acrylamide gels. Here we describe a novel HPLC-based method for identification and characterization of clonal TCR- targets. Clonality of PCR amplified TCR- products were examined on a high-resolution micropellicular DNASep™ matrix (WAVE® Nucleic Acid Fragment Analysis System, Transgenomic) and subsequently classified into clonal, biclonal or negative samples. V 2-D 3 rearrangements were analyzed by DHPLC, using triethylammonium acetate as an ion-pairing reagent, with a linear acetonitrile gradient at 50°C. Biclonal elution profiles consisted of two individual homoduplex peaks and one heteroduplex peak unique for each patient sample. For characterization of biclonal rearrangements DHPLC separated samples were subjected to a second run and individual peaks were collected. A software-controlled fragment collector was arranged in tandem with the HPLC system for this purpose and purified PCR products were collected in a time-dependent manner. Fractions were air dried and subsequently sequenced directly. The specificity of the observed patient specific sequences was tested by real time quantitative (RQ) PCR on a LightCycler system.

Chapter 8: SIRPH: a rapid, quantitative SNuPE-IP RP HPLC method for simultaneous methylation analysis at multiple CpG sites

Rapid mapping and quantification of DNA methylation is of high importance for the fields of developmental biology, cancer and gene imprinting. In this chapter, we describe in detail the SIRPH protocol for quantitative methylation analysis of PCR products derived from bisulfite treated DNA. The method is based on a combination of single nucleotide primer extension (SNuPE) and ion pair reversed phase high performance liquid chromatography (IP RP HPLC) separation techniques. Methylated and un-methylated CpG sites can be discriminated and quantified based on the different masses and hydrophobicities of the extended primer products. The assay is easy to perform, reproducible and several CpG sites can be analysed simultaneously in one reaction.

Chapter 9: Isolation and sequencing of heteroduplex fragments from DHPLC analysis of somatic, low-frequency mutations

DHPLC is a sensitive method for the detection of the presence of low-frequency mutations in biopsies from solid tumors. A reliable characterization of these low-frequency mutations by direct sequencing can be difficult. Here, a procedure for the automated isolation of small heteroduplex peaks, followed by re-amplification and sequencing is illustrated using two mutations in the p53 gene as examples. The sensitivity of DHPLC in detecting low-frequency mutations and the quality of sequencing data from isolated fractions depends on the quality of PCR products and their melting profiles. Optimase™ polymerase is shown to provide PCR products virtually free of polymerase-induced errors even after a nested PCR approach suitable for genomic DNA templates of variable quality. Finally, a procedure with a set of universal sequencing primers is described allowing consistent sequencing data to be obtained from very near the end of the amplicon-specific primers.

Chapter 10: Analysis of promoter methylation using denaturing high-performance liquid chromatography (DHPLC)

Cytosine methylation at CpG dinucleotides in promoter regions is an important control mechanism of tumor suppressor gene (TSG) down-regulation in cancer development and progression. DHPLC is a reliable, highly sensitive technique primarily developed for mutation detection. In this chapter DHPLC is described as a reliable pre-screening method for the detection of CpG methylation. The method for detection of methylation patterns by DHPLC was established by analyzing the methylation pattern of a 392-bp promoter fragment that contains 35 CpG sites and by using DNA samples with different, but well characterized, methylation patterns from either tumor specimens or from samples generated by the treatment of unmethylated blood DNA with DNA methylases. Bisulfite treatment of genomic DNA was employed for the deamination of unmethylated cytosine, resulting in their conversion to uracil. The promoter fragment was PCR-amplified with primers suitable for the amplification of unmethylated as well as methylated alleles. PCR products were denatured and reannealed, which results in the formation of heteroduplex DNA that is detectable by DHPLC. The optimal temperature at which heteroduplex detection occurred was deduced from the melting profile of the specific DNA fragments as predicted by WAVEMAKER™ software (Transgenomic). DHPLC analysis conditions were optimized using positive controls that were generated with specific DNA methylases. Complete methylation results in a peak shift, i.e. increased retention time of the analyzed PCR fragment compared to the PCR fragment derived from unmethylated blood DNA. Partial methylation of genomic DNA becomes apparent through the appearance of additional signals/peaks in the DHPLC trace that represent the formation of diverse heteroduplex structures. Thus, DHPLC is an appropriate high-throughput pre-screening technique for the detection of promoter methylation of putative tumor suppressor genes. The technique allows multiple CpG sites to be screened simultaneously for their methylation status. This is advantageous when analyzing genes with unknown methylation status. Verification and characterization of methylation status by subcloning and sequencing is only required for DHPLC-positive samples, which constitutes a significant reduction in analysis time.

Editor

Karl Heinz Hecker, Ph.D.

Karl Hecker is a Staff Scientist in the Separations & Analysis unit at Invitrogen™ Corporation in Carlsbad, California, focusing on the development of purification and analysis technologies for nucleic acids. Before joining Invitrogen in September 2002, he was Director of Applied Research at Transgenomic in San Jose, CA. He initially joined the Research & Development group at Transgenomic in San Jose in September 1998 as a Research Scientist. His areas of research involved separation, characterization, and purification of nucleic acids (DNA and RNA) using IP-RP HPLC/WAVE® System technology, the development of new applications for this emerging technology and the evaluation of new promising technologies. Prior to joining Transgenomic he was a postdoctoral fellow with Professor J. W. Streilein at the Schepens Eye Research Institute and Harvard Medical School, Department of Ophthalmology from 1995 - 1998 conducting research in ocular immunology and molecular biology. From 1993 - 1995 he held a postdoctoral position at the Department of Biological Sciences, Florida State University with Professor K. Roux working on variable region heavy chain genes. He received his Ph.D. in Biochemistry in 1993 from The Florida State University, Tallahassee working with Professor R. Rill on the interaction of small molecules/drugs with nucleic acids. Dr. Hecker received his MS degree in inorganic chemistry in 1986 from Eastern Illinois University working with Professor R. L. Keiter on phosphine trans-substituted iron pentacarbonyl complexes. He holds a BS (1984) and MS (1988) degree in chemistry from Johann Wolfgang Goethe University in Germany.

Contact
Karl Heinz Hecker, Ph.D.
Invitrogen Corporation
1600 Faraday Avenue
Carlsbad, CA 92008, USA
karl.hecker@invitrogen.com

Preface

Genetic variance detection, the discovery of and screening for single-nucleotide changes, short insertions, and deletions, is an integral part of genomic research and diagnostics. A highly sensitive and specific, reproducible, robust, and cost-effective technology for the detection of variance on the nucleic acids level amenable to automation, is pivotal to exploiting the intricate relationship of sequence variability and function. Many gel-based approaches for the detection of genetic variance with various degrees of sensitivity and specificity, but all labor intensive and with limited potential for automation, had been developed by the time when Denaturing High Performance Liquid Chromatography (DHPLC) emerged as an exquisitely sensitive and specific technology for genetic variance detection just a few years ago that was suitable for automation. I had the opportunity to become part of this exiting field in its early stages when I joined the R&D group at Transgenomic back in 1998, affording me the opportunity to contribute to and witness the proliferation and maturation of an exciting technology. This book contains contributions from a select group of scientists who have adopted the technology at various stages of its development and mastered and perfected its use for their specific purposes. In the process, these scientists have made significant contributions to the establishment, development, and growth of a new technology by defining its capabilities, demonstrating its proper use, and taking advantage of its potential. This book is intended to teach not only the novice, but also the advanced user of the technology by providing a detailed account of the fundamentals and a carefully chosen set of specific and innovative applications from the masters.

I am grateful to those who have generously contributed their valuable time, efforts, and material, sharing their expertise and knowledge of the technology. As the editor of this book, it is my privilege to have been associated with this publication and having had the opportunity to work together with its contributors. Many people have contributed along the way to making my involvement in this project possible and pleasurable. Thanks go out to Dr. Alexander Kuklin for offering me this editorship and for his continuous support and encouragement during the lengthy process of trying to keep everything moving along. I also wish to thank Alex Nartea for his patience and dedication in the formatting of the book for publication. Deep-felt thanks go out to my former colleagues and friends at Transgenomic for affording me the opportunity of participating and contributing to the emergence and establishment of an exceptional technology. I thank the many users of the technology for their feedback, questions, and comments over the years, which forced me to think about the technology in ways that deepened my understanding and facilitated continued application development and maturation of the technology.

Special thanks go out to my family and friends for their support and encouragement and to my wife Wanping and daughter Amanda for their love and affection.

Karl Heinz Hecker
San Diego, CA
June 2003

SECTION I

FUNDAMENTALS OF DHPLC

CHAPTER

1

BASICS OF AUTOMATED, HIGH-ACCURACY MUTATION SCREENING WITH THE WAVE® NUCLEIC ACID FRAGMENT ANALYSIS SYSTEMS

Karl H. Hecker, Ph.D.

Contact:
Karl Hecker, Ph.D.
Invitrogen Corporation
1600 Faraday Avenue, Carlsbad, CA 92008 U.S.A.
E-mail: karl.hecker@invitrogen.com

0-9664027-7-4/03/$0.00+$.50 *From:* **Genetic Variance Detection~Nuts & Bolts of DHPLC in Genomics** (pp.15-34)
Edited by: Karl H. Hecker, Ph.D.

1.1. Introduction

Studies of sequence variation are an integral part of genome research, because of differences in the genomic sequences between individuals. Variations in mammalian genomic sequences occur at an estimated frequency of at least one base per thousand bases. In the case of the human genome, this totals more than three million variances. These variances are typically present in the form of single nucleotide polymorphisms (SNP). The relationship between the variability in the sequence of specific genes among individuals to their disease status and/or susceptibility to disease is a primary key to understanding gene function. Variations, in addition to being critical for identification of disease genes, are also essential in understanding genetic differences in individual responses to the environment, to disease, and therapies. The challenge of understanding even a fraction of the variations in the genomic sequence, determining which of the polymorphisms are significant and which are inconsequential, is overwhelming. However, the potential value of such knowledge is great in that it will not only lead to better diagnostic and prognostic assays, but also to improved treatment for disease.

A highly sensitive, reproducible, and robust technology for detection of sequence variations is pivotal to exploiting the relationship between sequence variability and function. A recently developed technology for the detection of sequence variance, based on denaturing high-performance liquid chromatography (DHPLC)[1-3], for the analysis of DNA fragments or PCR products meets the above requirements. The technology, which is based on the detection of heteroduplexes, allows for automated, high-accuracy identification of SNP as well as small deletions or insertions. Heteroduplex profiles are easily distinguished from homoduplex peaks[3,4] and thus provide a reliable means for mutation scanning and discovery. DHPLC has been shown to clearly resolve mutations in various genes with detection rates ranging from 92.5% to 100%[5-8]. It is reported in the literature that the sensitivity of detection by DHPLC is higher than for alternative gel based analysis techniques, i.e. mutations not detected by gel-based techniques could be detected by DHPLC[8-10].

Here, principles underlying mutation detection by temperature modulated heteroduplex analysis (TMHA) using DHPLC technology on the WAVE® Nucleic Acid Fragment Analysis Systems from Transgenomic are explained and illustrated. Recent improvements in the WAVE System instrumentation and updates regarding the understanding of the technology are discussed. Furthermore, the importance of using WAVE Systems compatible samples and elution buffers and of using high-fidelity DNA polymerases for amplification of DNA fragments intended for analysis by DHPLC on the WAVE System are reviewed in the context of success, accuracy, and reliability of the WAVE System technology.

1.2. Methodology

1.2.1 WAVE Nucleic Acid Fragment Analysis Systems

All HPLC analyses shown here were performed on WAVE Nucleic Acid Fragment Analysis Systems equipped with DNASep® Cartridges (Transgenomic, San Jose, CA). Three major WAVE System configurations are currently available - 2100A, 3500A, and 3500 HT. In addition, the WAVE-MD™ System is now available. This system is based on Transgenomic's proven, highly sensitive DHPLC/TMHA technology with the DNASep Cartridge and is dedicated for mutation analysis. The WAVE-MD System employs fixed gradient conditions optimized for DNA fragments ranging in size from 100 - 600 base pairs. Other features include minimal set-up time, continuous and unattended operations, user-friendly WAVE-MD software, and cost effective consumables solutions with a targeted throughput of circa 100 samples per day. Key features of the individual system configurations are listed in ***Table 1-1***.

Chromatography on the WAVE Systems is performed with a dual buffer system, where buffers A is an aqueous solution of 0.1M triethylammonium acetate (TEAA) and buffer B a 0.1 M TEAA solution containing 25% acetonitrile (ACN). ACN (Transgenomic, San Jose, CA), TEAA (Transgenomic, San Jose, CA), and water used for the preparation of buffers were of HPLC grade. Volumetric flasks were used for the preparation of buffers due to the sensitivity of retention times to even slight variations in ACN concentrations. Alternatively, pre-made 1x buffers A and B (Transgenomic) can be used. Eluent conditions for analysis can be determined using either WAVEMAKER™ or WAVE Navigator™ software (Transgenomic, San Jose, CA). Eluent flow rates for all analyses were 0.9 mL/min in the Standard Analysis Mode and 1.5mL/min in the Rapid Analysis Mode. Chromatograms are generally recorded at a wavelength of 260nm or, in the case of fluorescently labeled DNA, at a wavelength that is characteristic for the fluorescent label used. All analyses under non-denaturing conditions are run at 50°C. Mutation detection under partially denaturing conditions is performed at or near the software predicted temperature indicated for each chromatogram.

1.2.2 DNASep Cartridge Technology

The DNASep Cartridge is the heart of the WAVE Nucleic Fragment Analysis System. The cartridge contains a robust, nonporous, alkylated poly(styrene-divinylbenzene) separation matrix[11-13]. Depending on the analysis mode, normal or rapid, a DNASep Cartridge with dimensions of 4.6 x 50mm or 6.5 x 37 mm is used. The diameter of the DNASep Cartridge for the WAVE System 3500HT in the Rapid Analysis Mode is increased in order to accommodate higher flow rates. The WAVE-MD System utilizes a 4 x 10mm separation cartridge, which is optimized for mutation detection under the standardized conditions of this system.

Under the analysis conditions employed for mutation detection with the WAVE System, the separation matrix exhibits a higher affinity for double-stranded DNA (dsDNA) than for single-stranded DNA (ssDNA). Mutation detection is performed at or slightly below the melting temperature of the homoduplex. At this temperature heteroduplexes, if present, are partially melted at the mismatch site and therefore exhibit partial ssDNA character. This results in a decreased retention time of partially denatured heteroduplexes as compared to the corresponding homoduplexes.

1.2.3 WAVEMAKER and WAVE Navigator Software

WAVEMAKER and WAVE Navigator software packages control the operation of the HPLC hardware. Melt profiles of DNA fragments being analyzed for the presence of mutations, analysis temperatures, and HPLC gradient conditions are predicted by WAVEMAKER or WAVE Navigator software (Transgenomic, San Jose, CA). The analysis temperatures for mutation detection are derived from software predicted melt profiles of percent helical fraction versus base position. Both software packages employ a modified Fixman-Friere algorithm to predict the sequence-dependent melt behavior of DNA fragments[14,15]. A temperature at which the target fragment, the PCR product being scanned for the presence of mutations, is 70-85% helical is regarded as the optimal temperature for mutation detection. Both software packages control the acquisition and handling of data. Data analysis features vary depending on the software package used. WAVE Navigator is the latest software product for the operation of the WAVE System and analysis of chromatographic results. It includes enhanced features over the WAVEMAKER software such as a relational database management system, automated mutation calling, more robust and expansive data storage capabilities, across project data analysis, and remote operation capabilities.

1.2.4 Sample Preparation

DNA fragments to be analyzed for the presence of mutations are most commonly PCR products, but could also be restriction fragments. Primers flanking the target region for mutation detection are generally designed using conventional primer design software packages such as OLIGO® (Molecular Biology Insights Inc., Cascade, CO) or Primer 3 (http://www-genome.wi.mit.edu/cgi-bin/primer/primer3_www.cgi). Transgenomic highly recommends that each amplicon sequence be evaluated for its suitability to detect mutations by heteroduplex analysis on the WAVE System using either WAVEMAKER or WAVE

Navigator software. Additionally, a tool for amplicon design is offered at www.mutationdiscovery.com for subscribers of an instrument maintenance plan. PCR products amplified from heterozygous templates generate wild type and mutant allele products in equimolar amounts. PCR products from homozygous templates require post-PCR addition of wild type allele PCR product in order for mutations to be detected. Prior to mutation analysis PCR products need to be denatured and slowly reannealed to promote the formation of heteroduplexes. The latter is generally achieved by heating to 95°C, followed by slow cooling of the sample to 25°C at a rate of minus 0.1°C per minute. PCR products and restriction digests do not need to be purified prior to mutation analysis on the WAVE System. In fact, purification of DNA fragments prior to analysis on the WAVE System is discouraged, because of a possible introduction of compounds from the purification kit or procedure that may be harmful to the performance and durability of separation cartridge.

1.3 Results and Discussion

1.3.1 Principles of Mutation Detection by TMHA

The screening and scanning for sequence variation/mutations by DHPLC involves the analysis of homoduplexes and heteroduplexes that are formed between DNA fragments (PCR products or restriction fragments) derived from wild type and mutant sequences (Figure 1-1). In brief, a primer pair is designed or restriction sites are chosen to flank the DNA sequence of interest. DNA fragments ranging in size from ca. 150 to 1,000bp or longer may be suitable for analysis. The decisive factors in judging whether a DNA fragment is suitable for mutation detection are its length, sequence, and melt characteristics[16]. PCR or endonuclease digestion of heterozygous templates generates wild type and mutant allele products in equimolar amounts. DNA fragments generated from homozygous mutant template require the addition of equimolar amounts of wild type product to permit formation of heteroduplexes. Wild type product is generally added when screening for unknown mutations and for the analysis of samples when the occurrence of homozygous mutants is likely or possible. This approach, in the case where the DNA fragment is generated by PCR, is termed mutation detection after post-PCR mixing. A wild type/mutant mixture is denatured at 95°C. This is followed by slow renaturation, which leads to the formation of wild type and mutant homoduplexes as well as two distinct wild type/mutant heteroduplexes. The latter are characterized by a mismatch at the site where the sequence variance occurs.

Mutation analysis by DHPLC on the WAVE System, which employs the technique of Temperature Modulated Heteroduplex Analysis (TMHA), is based on two key features. First, heteroduplexes always melt at the mismatch site generat-

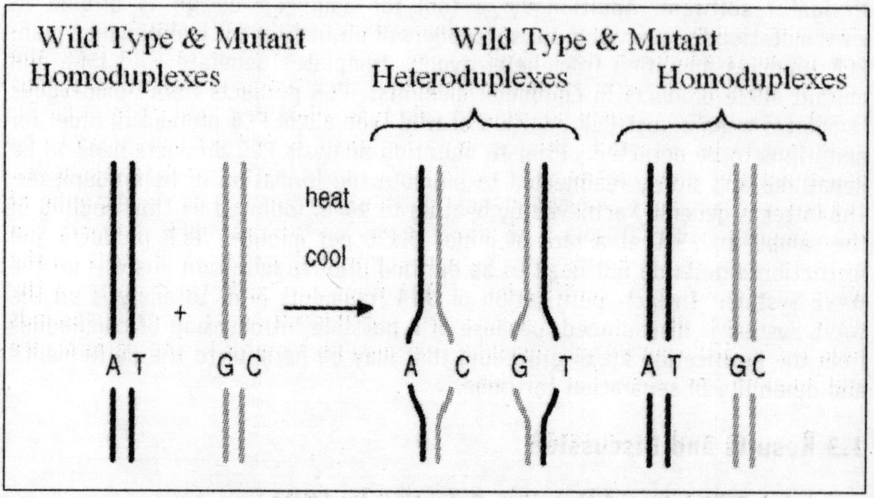

Figure 1-1. Schematic of duplex formation. Mixtures of wild type and mutant amplicons, which are either co-amplified during the PCR or generated by post-PCR mixing, are denatured by heating and reannealed by slow cooling, leading to the formation of four distinct species, two original homo- and two heteroduplexes. Sequence variations resulting in heteroduplex formation include SNP as well as short insertions and deletions. Wild type and mutant homoduplexes melt at higher temperatures than mismatch containing wild type/mutant heteroduplexes. Differences in the melting temperatures are the basis for the identification of mutations by TMHA.

ing a single-stranded, partially denatured region at a temperature below the melting temperature of the corresponding wild type and mutant homoduplexes. Second, ssDNA is retained less strongly on the DNASep Cartridge than dsDNA. Thus, at a temperature at which wild type and mutant homoduplexes are still partially denatured, resulting in up to two distinct peaks, the wild type/mutant heteroduplexes are already denatured to a greater extent and are less well retained on the column. The more denatured heteroduplexes yield up to two peaks of equal intensity preceding the homoduplex peak(s).

The DNASep Cartridge contains a nonporous, C18-alkylated, and therefore hydrophobic poly(styrene-divinylbenzene) separation matrix. This precludes direct interactions between the hydrophilic nucleic acids analyzed and the hydrophobic separation matrix. Interactions between nucleic acids and the separation matrix are mediated by triethylammonium acetate (TEAA), which is contained in both buffers A and B at a concentration of 100 mM. Upon sample injection, the cationic triethylammonium ions interact electrostatically, e.g. bind non-covalently, to the anionic nucleic acid phosphate backbone. The alkyl groups of the triethylammonium cations interact with the hydrophobic surface of the alkylated DNASep matrix, mediating the interaction between nucleic acids and the separation matrix.

Eluent buffer A is a purely aqueous solution that is 100 mM in TEAA. Eluent buffer B is 100 mM in TEAA and contains 25 % (v/v) of the organic solvent acetonitrile (ACN). The DNASep Cartridge is charged with a nucleic acids sample at a software-predicted ratio of buffers A and B, which ensures binding of the DNA fragments to the column matrix. Elution of the nucleic acid fragments analyzed, in a manner that is dependent on their lengths as well as their state of naturation/denaturation, is then achieved by a systematic, software-predicted increase in elution buffer B relative to A. Under gradient conditions used for mutation detection nucleotides and primers used in PCR exhibit virtually no affinity for the DNASep matrix. Consequently, these nucleic acids elute early on in the elution gradient, evidenced by the presence of a large peak in the chromatogram preceding the analysis portion of the gradient (Figure 1-2, peak at 0.5 - 1.0 min). Polymerases and endonucleases, or proteins at low concentrations in general, are not retained on the separation matrix under the analysis conditions used for mutation detection. Proteins contained in reaction mixtures are therefore co-eluted with nucleotides and primers at the beginning of the elution gradient. Notable exceptions are extremely hydrophobic proteins, e.g. bovine serum albumin (BSA), at high concentrations, which potentially accumulate on the column matrix and interfere with column performance, retention times as well as resolution. The elution of the nucleic acids is monitored on-line by either UV or, in the case of fluorescently labeled nucleic acids, by fluorescence detection.

1.3.2 Identifying the Presence of a Genetic Variation - Mutation Calling

Single nucleotide changes as well as short insertions and deletions can be detected with high accuracy and sensitivity by TMHA using either one of the WAVE Nucleic Acid Fragment Analysis Systems or the WAVE-MD System. Sequence variations are identified based on a comparison between the chromatograms of wild type amplicons, e.g. devoid of sequence variations, and chromatograms of amplicons that carry mutations, which have formed heteroduplexes with wild type amplicons, recorded under identical chromatographic conditions.

A characteristic example of mutation calling is presented in Figure 1-2, A - C. The wild type chromatogram exhibits a typical single-peak pattern, because the amplified wild type sample contains a single homoduplex amplicon. With increasing analysis temperature the PCR product progressively melts, resulting in a steady increase in its single-stranded character and causing a steady reduction in retention time. The latter is the result of decreased affinity of ssDNA to the column matrix as compared to dsDNA. The single-peak chromatogram of a wild type amplicon, recorded at or near the software-predicted melting temperature of the wild type homoduplex, serves as a reference for mutation calling.

The presence of sequence variations/mutations in an amplified region leads to the formation of four distinct duplexes (Figure 1-1) during the denaturation and reannealing step preceding TMHA. In the case of the wild type/mutant 1 chromatogram (Figure 1-2B), where the mutant is defined as having a C→T change in position 178 of the amplified region, the presence of a sequence variation becomes apparent by the appearance of additional peaks in the chromatograms. In contrast to the chromatograms for the wild type amplicon, where only one homoduplex is present in the sample (Fig. 1-2A), the 1:1 wild type/mutant 1 mixture contains four distinct duplexes, two homoduplexes that correspond to one wild type and one mutant homoduplex and two heteroduplexes, which are formed by the (+/-) and the (-/+) strands of the wild type and mutant allele PCR products. Homoduplexes have higher melt temperatures than the corresponding mismatch containing heteroduplexes. Thus, the retention times of peaks representing homoduplexes is always longer than retention times of their heteroduplex counterparts under identical gradient conditions and at analysis temperatures at or near the melt temperature of the homoduplex.

In the wild type/mutant 1 chromatograms presented here, the two heteroduplex species are not resolved from each other at any of the temperatures

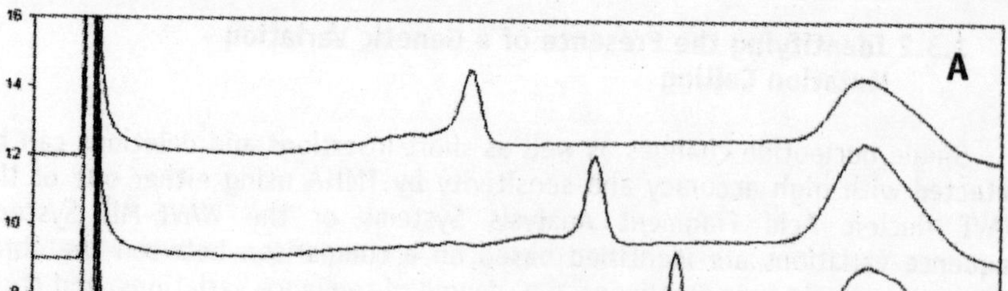

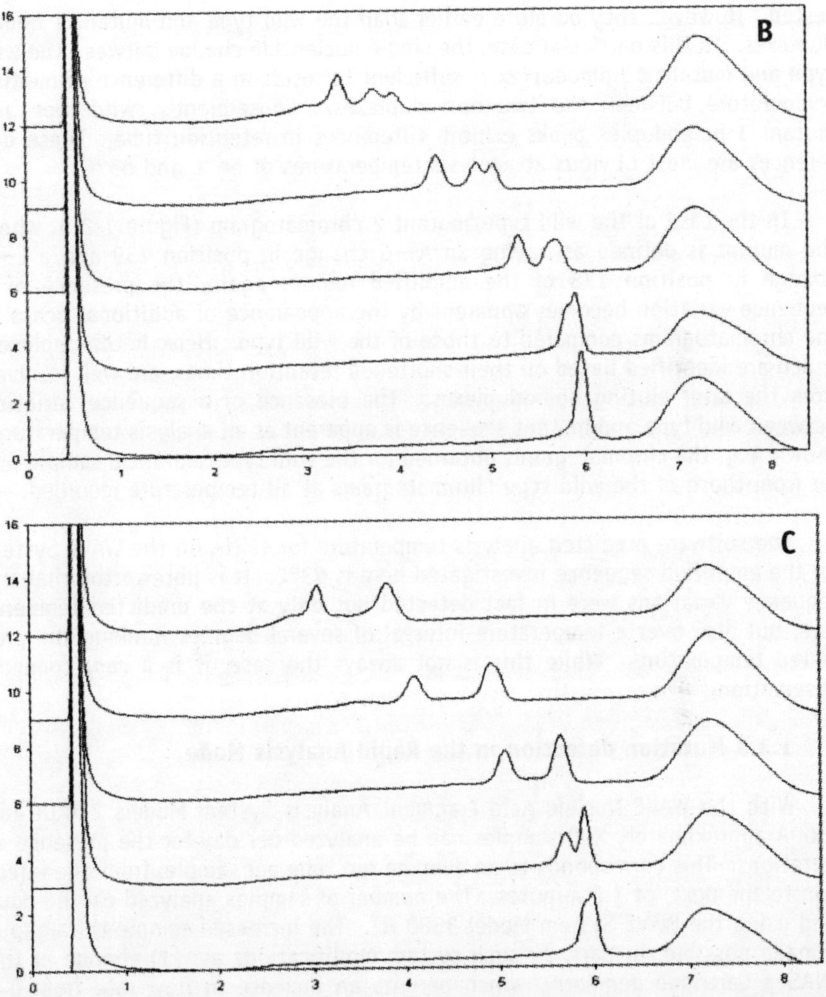

Figure 1-2. Mutation calling. Mutation analysis by TMHA was performed on 304bp-long PCR products amplified from plasmid clones A (wild type), B (Mutant 1: C→T at position 178), and C→Mutant 2: A→G at position 139 and C→T at position 178). PCR products from wild type and mutant templates were mixed in a 1:1 ratio, denatured, and slowly reannealed to form mixtures of hetero- and homoduplexes. All chromatograms were recorded under identical gradient conditions. TMHA was performed over a temperature range from 61 to 65°C. The bottom chromatograms in panels A - C correspond to an analysis temperature of 61°C and the top chromatograms to an analysis temperature of 65°C.

tested. However, they do elute earlier than the wild type and mutant 1 homoduplexes. In this particular case, the single nucleotide change between the wild type and mutant 1 homoduplex is sufficient to result in a difference of melting temperature between the two homoduplexes. Consequently, wild type and mutant 1 homoduplex peaks exhibit differences in retention time. These differences are most obvious at analysis temperatures of 64°C and 65°C.

In the case of the wild type/mutant 2 chromatogram (Figure 1-2C), where the mutant is defined as having an A→G change in position 139 and a C→T change in position 178 of the amplified region, again, the presence of a sequence variation becomes apparent by the appearance of additional peaks in the chromatograms compared to those of the wild type. Here, heteroduplexes, which are identified based on their shortened retention times, are well resolved from the later eluting homoduplexes. The presence of a sequence variation between wild type and mutant sequence is apparent at all analysis temperatures shown, e.g. the chromatograms obtained for the wild type/mutant 2 sample differ from those of the wild type chromatograms at all temperature recorded.

The software predicted analysis temperature for TMHA on the WAVE System for the amplified sequence investigated here is 63°C. It is noteworthy that the sequence variations were in fact detected not only at the predicted temperature, but also over a temperature interval of several degrees flanking the predicted temperature. While this is not always the case, it is a very common observation[17,18].

1.3.3 Mutation detection in the Rapid Analysis Mode

With the WAVE Nucleic Acid Fragment Analysis System Models 2100A and 3500A approximately 200 samples can be analyzed per day for the presence of mutations. This corresponds to an average run time per sample, from one injection to the next, of 7.2 minutes. The number of samples analyzed can be doubled using the WAVE System Model 3500 HT. The increased sample throughput is made possible, in part, by such system modifications as: (1) change of the DNASep Cartridge geometry, which permits an increase in flow rate from 0.9 mL/min to 1.5 mL/min, (2) use of the smaller eluent mixer, 0.250 mL as compared of 2.00 mL, which supports faster and more precise changes in eluent composition, and (3) the use of an accelerator, which allows for an accelerated cartridge-cleaning step as well as reduced equilibration times between injections (see ***Table 1-1***). Two examples of mutation detection with the WAVE System 3500 HT in the Rapid Analysis Mode, using Transgenomic's mutation standards for 56°C and 63°C, are shown in Figure 1-3.

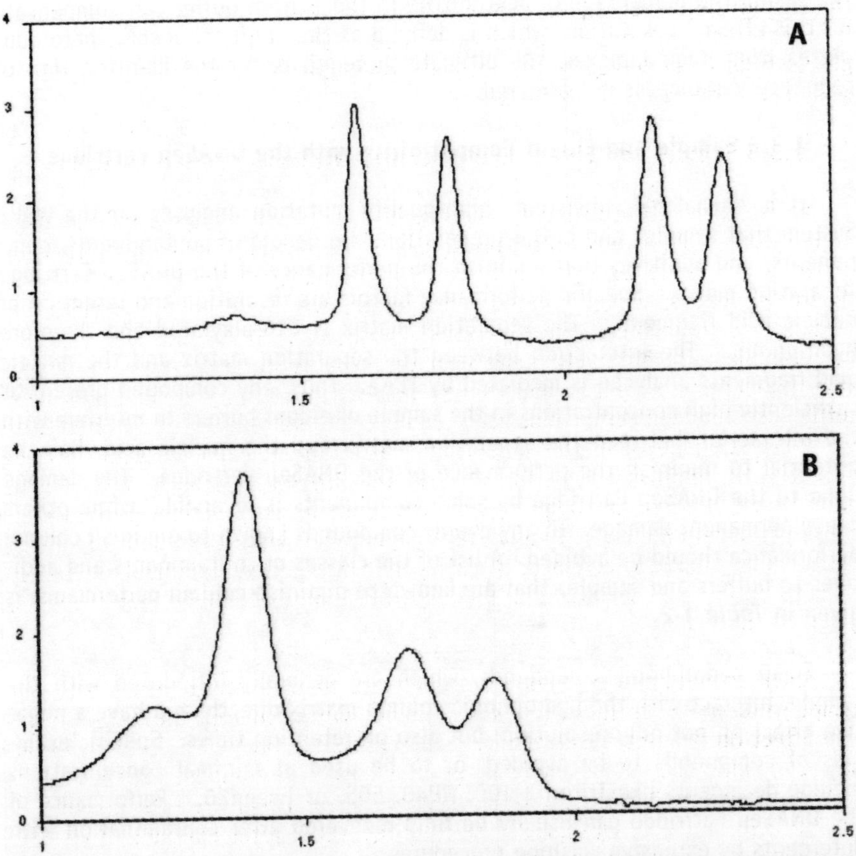

Figure 1-3. Mutation detection in Rapid Analysis Mode. Mutation detection with the WAVE System 3500 HT using the Rapid Analysis Mode is demonstrated here with two mutation standards (Transgenomic) at 56°C (Panel A) and 63°C (Panel B). Only the analysis portion of the run from 1 - 2.5 minutes is shown.

1.3.4 Mutation Detection Using Fluorescently Labeled DNA Fragments

The use of fluorescent labels is compatible with the technology of mutation detection by DHPLC on the WAVE System and generally increases sensitivity[19,20]. The affinity of the column matrix for the labeled DNA fragment is increased due to the presence of hydrophobic fluorescent dye molecules.

Increasing the percentage of acetonitrile in the elution buffer can compensate for this effect. Resolution, which is defined as the ability to resolve heteroduplexes from homoduplexes, the ultimate prerequisite for the identification of sequence variance, is not affected.

1.3.5 Sample and Eluent Compatibility with the DNASep Cartridge

It is crucial for consistent, high-quality mutation analyses on the WAVE System that samples and buffer preparations are devoid of contaminants, components, and additives that diminish the performance of the DNASep Cartridge separation matrix. Specific performance factors are resolution and retention of nucleic acid fragments. The separation matrix is C18-alkylated and therefore hydrophobic. The interaction between the separation matrix and the nucleic acid fragments analyzed is mediated by TEAA. Thus, any compound present at sufficiently high concentrations in the sample or eluent buffers to interfere with the interaction between the separation matrix and the nucleic acid, has the potential to diminish the performance of the DNASep Cartridge. The damage done to the DNASep Cartridge by some components is reversible, while others cause permanent damage. In any event, compounds known to diminish column performance should be avoided. A list of the classes of contaminants and additives to buffers and samples that are known to diminish column performance is given in ***Table 1-2***.

Small hydrophobic compounds, which are generally introduced with the sample, interact with the hydrophobic column matrix directly and have a negative affect on not only resolution, but also on retention times. Specific examples of compounds to be avoided, or to be used at minimal concentrations include detergents like Triton X-100, NP40, SDS, or Tween20. Performance of the DNASep Cartridge can usually be fully recovered after contamination with detergents by extensive washing procedures.

Aromatic and aliphatic compounds, which are hydrophobic, can dramatically interfere with column performance. Phenol, which can be carried over after phenol/chloroform extraction of template DNA, does not only interfere with PCR, but also negatively affects DNASep cartridge performance. Mineral oil dramatically and irreversibly alters the performance of the DNASep Cartridge. It is not to be used as an overlay in PCR reactions intended for analysis on the WAVE System.

Certain, mainly polyvalent metal ions, but not Mg^{2+}, will interfere with column performance, e.g. causing severe loss of resolution and variable retention times, because of their ability to bind to the negative phosphate backbone of the nucleic acids analyzed. Sources of polyvalent cation contamination can be poor quality of the water used for buffer preparation, glassware, or prolonged contact of the eluent buffers with metal parts. The latter can lead to iron contamination, which has a particularly severe effect on column performance.

Macromolecules and particulate material can physically interfere with column performance by obstructing eluent flow and preventing interactions between the column matrix and the nucleic acid fragments analyzed. Macromolecules, such as large, particularly denatured proteins and carbohydrates, can be carried over with the template or may have been added as enhancers to the PCR, e.g. polyethylene glycol (PEG) or heparin. It is essential to remove macromolecules and particulate material from samples prior to analysis on the WAVE System or to avoid their introduction in the first place.

Several chemicals commonly used as PCR additives and enhancers, such as formamide, glycerol, DMSO, and betaine, should be avoided or at least used with caution, because of their potential to interfere with column performance and/or chemically modify the nucleic acids analyzed. In the case that the use of the aforementioned additives is considered necessary, it is recommended to closely monitor column performance by frequently running mutation standards.

The WAVE System with its DNASep Cartridge provides a robust and reliable technology for the detection of mutations. Consideration of the aforementioned factors affecting column performance will not only ensure consistent, high-quality results, but also long cartridge life.

1.3.6 Importance of Polymerase Fidelity for Mutation Detection

DNA polymerases inherently incorporate errors randomly during amplification[21-23]. The extent of error incorporation depends in part on a number of factors, including e.g. dNTP and Mg^{2+} concentration. However, the primary factor determining the error incorporation rate is the polymerase itself. Polymerases that have 3′ to 5′ exonuclease activity, generally referred to as proofreading activity, incorporate the least number of errors and exhibit highest replication fidelity[24-27]. It is apparent that errors incorporated at random by the polymerase during amplification have the potential to interfere with the identification of mutations present in the amplified template. Therefore, the use of a high fidelity polymerase such as Transgenomic's Optimase™ Polymerase is recommended for the amplification of PCR products to be used for mutation detection.

Representative data on the outcome of different error incorporation rates by several commonly used DNA polymerases and their influence on the ability to reliably call mutations is shown in Figure 1-4. A 500-bp PCR product was

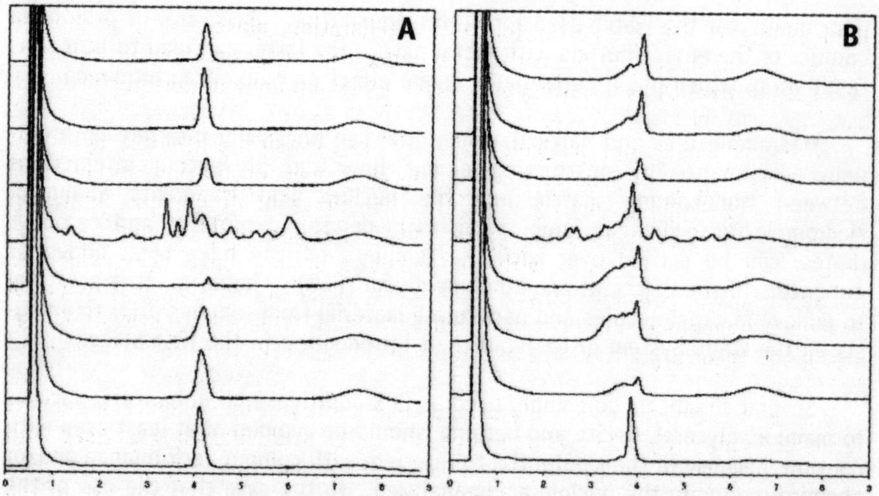

Figure 1-4. Importance of polymerase fidelity. A 500-bp fragment was amplified from phage genomic DNA with several different polymerases with and without proofreading activity. Transgenomic's Optimase® Polymerase with proofreading activity was one of the polymerases tested. Optimase Polymerase results are shown in the bottom chromatograms of panels A and B. Due to the high yield obtained with Optimase Polymerase, chromatograms of PCR products for this polymerase are scaled by a factor of 0.5 in panel A and 0.3 in panel B. PCR buffers used were those provided with each of the polymerases tested. All amplifications were performed under identical cycling conditions. Panel A shows PCR products analyzed under nondenaturing conditions for assessment of PCR product quality and yield. The fidelity of polymerization was assessed at 62°C, which is the software predicted TMHA temperature of the amplified DNA fragment(panel B). Errors caused by polymerase infidelity lead to the formation of many different heteroduplexes. Heteroduplexes elute earlier than homoduplexes and are apparent as broad peaks preceding the homoduplex peak.

amplified from phage genomic DNA using several different polymerases. PCR products were first analyzed under nondenaturing conditions on the WAVE System at 50°C in order to assess PCR product quality and yield. Results are shown in Figure 1-4, Panel A. As expected, product yield and quality varied considerably for the different polymerases tested. The intensity of the product peak is proportional to PCR product yield. Peak intensities reveal yields ranging from poor to good. PCR quality is more difficult to evaluate under nondenaturing conditions than yield. In one case, however, product quality was apparently not acceptable as evidenced by the presence of multiple spurious peaks, with the target product accounting for only a small portion of the total amount of PCR product amplified. In other cases, only a lack of symmetry of the product peak gives a first indication of potentially poor PCR product quality.

A true measure of PCR product quality is obtained when analyzing amplified PCR products at the software predicted temperature for TMHA and under the software predicted gradient conditions. PCR products that contain polymerase-induced errors will form heteroduplexes with the PCR products of the correct sequence. As seen above, heteroduplexes will elute earlier than their corresponding homoduplexes. Therefore, the extent of UV absorbance recorded preceding the main homoduplex peak is indicative and directly proportional to the amount of PCR product present that carries errors introduced by the polymerase used. In the examples shown in Figure 1-4B all polymerases but Transgenomic's Optimase™ Polymerase show significant error incorporation that will interfere with mutation calling. Heteroduplexes formed due to the presence of sequence variations in the template will co-elute with those heteroduplexes formed due to polymerase-induced errors. As the fidelity of the polymerase used decreases, formation of heteroduplexes resulting from PCR products carrying polymerase-induced errors will increase. As a consequence, accurate and reliable identification of true sequence variations will become increasingly difficult.

In order to maintain high sensitivity and accuracy for mutation detection by DHPLC, or for any other technology for mutation detection, the use of high-fidelity DNA polymerases and PCR conditions promoting polymerase fidelity is essential. Transgenomic recommends the use of Optimase™ Polymerase for amplification of DNA fragments intended for mutation analysis with the WAVE System. Further discussions on the use of proof-reading polymerases are given in Chapter 2 of this book.

1.4 Conclusion

All WAVE Nucleic Acid Fragment Analysis System configurations (2100A, 3500A, 3500HT, and WAVE-MD) provide rapid and high throughput DNA fragment separations on DNASep polymeric cartridges for the purpose of detecting genetic variations such as single nucleotide substitutions, mutations or polymorphisms, as well as short insertions and deletions. Sensitivity and specificity of genetic variance detection with the WAVE System consistently outperforms that of other technologies, including sequencing, at a fraction of their cost and/or labor involved[28]. The WAVE System provides a highly sensitive, efficient, and inexpensive technology platform for detecting genetic variation. UV and fluorescence on-line detection capabilities afford experimental flexibility. User-friendly software for the prediction of gradient and analysis conditions and fully automated sample handling, analysis, and data acquisition contribute to the high degree of automation and ease of use of the WAVE System technology.

Table 1-1: WAVE® Systems and Key Features

Feature	2100A	3500A	3500 HT	WAVE-MD™
Column	4.6 x 50mm DNASep Cartridge	4.6 x 50mm DNASep Cartridge	6.5 x 37mm DNASep Cartridge	4 x 10mm DNASep MD Cartridge
Accelerator	None	Internal	Internal	None
Oven Calibration Accuracy	Multi Point 0.1°C	Multi Point 0.1°C	Multi Point 0.1°C	Multi Point 0.1°C
Mixer	2 mL	2 mL	0.250 mL	High Pressure Mixing
Flow Rate	0.9 mL/min	0.9 mL/min	0.9 mL/min (normal) 1.5 mL/min (rapid)	0.36 mL/min
Throughput	±200 samples/day	±200 samples/day	±400 samples/day	±100 samples/day

WAVEMAKER software or WAVE Navigator software can be run with system configurations 2100A, 3500A, and 3500 HT. WAVE-MD Navigator™ software controls the operation of the WAVE-MD System. UV detector is standard for all systems. Fluorescence detector and fragment collector are optional for WAVE Systems 2100A, 3500A, and 3500 HT.

Table 1-2: Compound Classes Known to Affect DNASep Cartridge Performance

Compound	Class Examples
Hydrophobic Molecules	Detergents (Triton X-100, NP40, Tween20, SDS, etc.) Aromatic molecules (e.g. phenol) Aliphatic molecules (e.g. mineral oil)
Metal Cations	Polyvalent metal cations, e.g. Fe^{3+} (but not Mg^{2+})
Macromolecules & Particulate Material	Large and/or hydrophobic proteins, carbohydrates, polyethylene glycol, etc. Particulates
PCR Additives	Formamide, glycerol, DMSO, betaine

ACKNOWLEDGMENTS

The author thanks John Dektar for supplying chromatograms demonstrating the use of the WAVE System 3500 HT in Rapid Analysis Mode, Stacy Green for generating data and providing chromatograms shown in Figures 1-2 and 1-4, Tami Schmitt and the Transgenomic UK Optimase® Polymerase team for feedback and input regarding sample and eluent compatibility with the DNASep Cartridge and on the importance of polymerase fidelity for mutation detection analyses.

References

1. ***Oefner PJ, Underhill PA:*** Comparative DNA sequencing by denaturing high-performance liquid chromatography (DHPLC). *Am J Hum Genet 1995*; 57:A266.

2. ***Oefner PJ, Underhill PA:*** DNA mutation detection using denaturing high-performance liquid chromatography (DHPLC); in Current Protocols in Human Genetics. Wiley & Sons, New York, 1998, *Supplement 19*, pp 7.10.1-7.10.12.

3. ***Kuklin A, Munson K, Gjerde D, Haefele R, Taylor P:*** Detection of single-nucleotide polymorphisms with the WAVE™ DNA Fragment Analysis System. *Genetic Testing 1997/98*; 1:201-206.

4. ***Underhill PA, Jin L, Lin AA, Mehdi SQ, Jenkins T, Vollrath D, Davis RW, Cavalli-Sforza LL, Oefner PJ:*** Detection of numerous Y chromosome bialleleic polymorphisms by denaturing high-performance liquid chromatography. *Genome Res 1997*; 7:996-1005.

5. ***Liu W, Smith DI, Rechtzigel KJ, Thibodeau SN, James CD:*** Denaturing high performance liquid chromatography (DHPLC) used in the detection of germline and somatic mutations. *Nucl Acids Res 1998*; 26:1396-1400.

6. ***O'Donovan MC, Oefner PJ, Roberts SC, Austin J, Hoogendoorn B, Guy C, Speight G, Upadhyaya M, Sommer SS, McGuffin P:*** Blind analysis of denaturing high-performance liquid chromatography as a tool for mutation detection. *Genomics 1998*; 52:44-49.

7. ***Jones AC, Austin J, Hansen N, Hoogendoorn B, Oefner P, Cheadle JP, O'Donovan MC:*** Optimal temperature selection for mutation detection by denaturing HPLC and comparison to single-stranded conformation polymorphism and heteroduplex analysis. *Clin Chem 1999*; 45:1133-1140.

8. ***Choy YS, Dabora SL, Hall F, Ramesh V, Niida Y, Franz D, Kasprzyk-Obara J, Reeve MP, Kwiatkowski DJ:*** Superiority of denaturing high performance liquid chromatography over single-stranded conformation and conformation-sensitive gel electrophoresis for mutation detection in TSC2. *Ann Hum Genet 1999*; 63:383-391.

9. ***Gross E, Arnold N, Goette J, Schwarz-Boeger U, Kiechle M:*** A comparison of BRCA1 mutation analysis by direct sequencing, SSCP and DHPLC. *Hum Genet 1999*; 105:72-78.

10. ***Wagner T, Stoppa-Lyonnet D, Fleischmann E, Muhr D, Pages S, Sandberg T, Caux V, Moeslinger R, Langbauer G, Borg A, Oefner P:*** Denaturing high-performance liquid chromatography detects reliably BRCA1 and BRCA2 mutations. *Genomics 1999*; 62:369-376.

11. ***Huber CG, Oefner PJ, Bonn GK:*** Rapid and accurate sizing of DNA fragments by ion-pair chromatography on alkylated nonporous poly(styrene-divinylbenzene). *Analytical Chemistry 1995*; 67:578-585.

12. ***Huber CG, Oefner PJ, Preuss E, Bonn GK:*** High-resolution liquid chromatography of DNA fragments on non-porous poly(styrene-divinylbenzene) particles. *Nucl Acids Res 1993*; 21:1061-1066.

13. ***Huber CG, Oefner PJ, Bonn GK:*** Rapid analysis of biopolymers on modified non-porous poly(styrene-divinylbenzene) particles. *Chromatographia 1993*; 37:653-658.

14. ***Poland D:*** Recursion relation generation of probability profiles for specific-sequence macromolecules with long-range correlations. *Biopolymers 1974*; 18:1859-1871.

15. ***Fixman M, Freire JJ:*** Theory of DNA melting curves. *Biopolymers 1977*; 16:2693-2704.

16. ***Hecker KH, Asea A, Kobayashi K, Green S, Tang D, Calderwood SK:*** Mutation detection in the human HSP70B′ gene by denaturing high-performance liquid chromatography. *Cell Stress & Chaperones 2000*; 5(5):415-424.

17. ***Takashima H, Boerkoel CF, Lupski JR:*** Screening for mutations in a genetically heterogeneous disorder: DHPLC versus DNA sequence for mutation detection in multiple genes causing Charcot-Marie-Tooth neuropathy. *Genetics in Medicine 2001*; 3(5):335-342.

18. ***Spiegelman JI, Mindrinos MN, Oefner PJ:*** High-Accuracy DNA Sequence Variation Screening by DHPLC. *BioTechniques 2000*; 29(5):1084-1092.

19. ***Hecker KH, Taylor PD, Gjerde DT:*** Mutation Detection by Denaturing DNA Chromatography Using Fluorescently Labeled Polymerase Chain Reaction Products. *Anal Biochem 1999*; 272:156-164.

20. ***Oefner PJ, Huber CG, Umlauft F, Berti GN, Stimpfl E, and Bonn GK:*** High-resolution liquid chromatography of fluorescent dye-labeled nucleic acids. *Anal Biochem 1994*; 223:39-46.

21. ***Cline J, Braman JC, Hogrefe HH:*** PCR fidelity of Pfu DNA polymerase and other thermostable DNA polymerases. *Nucl Acids Res 1996*; 24:3546-3551.

22. ***Mattila P, Korpela J, Tenkanen T, Pitkanen K:*** Fidelity of DNA synthesis by the Thermococcus litoralis DNA polymerase - an extremely heat stable enzyme with proofreading activity. *Nucl Acids Res 1991*; 19:4967-4973.

23. ***Cha RS, Thilly WG:*** Specificity, efficiency and fidelity of PCR. *PCR Methods and Applications 1993*; 3:S18-S29.

24. ***Bessman MJ, Muzycyka N, Goodman MF, Schnaar RL:*** Studies of the biochemical basis of spontaneous mutation. II. The incorporation of a base and its analogue into DNA by wild type, mutator and antimutator polymerases. *J Biol Chem 1974*; 88:409-421.

25. ***Clayton LK, Goodman MF, Branscomb EW, Galas DJ:*** Error induction and correction by mutant and wild type T4 DNA polymerase: Kinetic error discrimination mechanisms. *J Biol Chem 1979*; 254:1902-1912.

26. ***Goodman MF, Fygenson DK:*** DNA polymerase fidelity: From genetics to a biochemical understanding. *Genetics 1998*; 148:1475-1482.

27. ***Reha-Krantz L:*** Regulation of DNA polymerase exonucleolytic proofreading activity: Studies of T4 "antimutator" DNA polymerases. *Genetics 1998*; 148:1551-1557.

28. ***Xiao W, Oefner PJ:*** Denaturing High-Performance Liquid Chromatography: A Review. *Human Mutation 2001*; 17:439-474.

CHAPTER

2

INFLUENCE OF PCR ON DHPLC POLYMORPHISM CHARACTERIZATION

Michael L. Nickerson

Contact:
Michael L. Nickerson
Laboratory of Immunobiology, National Cancer Institute–Frederick,
Building 560, Room 12-69, Frederick, MD 21702
E-mail: nickersm@mail.ncifcrf.gov

0-9664027-7-4/03/$0.00+$.50 *From:* **Genetic Variance Detection~Nuts & Bolts of DHPLC in Genomics** (pp.35-50)
Edited by: Karl H. Hecker, Ph.D.

2.1 Introduction

Denaturing high performance liquid chromatography (DHPLC) has emerged as a powerful tool to detect polymorphisms and mutations in genomes of living organisms[1-3]. Relative to other mutation detection techniques such as sequencing, primer extension, and SSCP, DHPLC is:

- **Sensitive:** clear separation between signal and noise (background) and polymorphisms in mixed or pooled samples can be visualized.

- **Versatile:** standard run parameters allow detection of single nucleotide polymorphisms, insertions, deletions, and more complex genetic changes.

- **Fast:** typically, a sample can be analyzed in shorter than 9 minutes with immediate accessibility to the data (chromatogram).

- **Economical:** after purchase of the HPLC apparatus, columns now have a long lifetime of several thousand runs (issues of polymerase reaction mixture components degrading column performance have been addressed, see discussion below), and no other specialized reagents are needed.

Other sections of the book cover these issues in greater detail but it is important to note that the first two points above are directly dependent upon the quality of the amplified PCR product that is being analyzed. Fortunately, reagent systems produced by a variety of companies have been optimized making amplification of 'clean' products (described below) relatively straightforward.

The most important element of PCR is the choice of polymerase and there is an amazing variety of enzymes available for amplification[4-6]. These range from high to low fidelity and possess complex reaction and storage buffers. The polymerase chosen for amplification of samples to be scanned for polymorphisms can be crucial to the quality of DHPLC signatures. Unfortunately, even though many authors describe PCR conditions in detail, including the polymerase used, a significant number of publications omit this information, or only refer the reader to additional references[7-11]. Additionally, only a few groups have directly examined how PCR conditions, especially choice of polymerase, affect DHPLC[12,13].

Some PCR-related variables have been extensively described, such as detergents and other components in polymerase reaction buffers, which affect DHPLC profiles and considerably shorten the lifetime of alkylated poly(styrene divinylbenzene) columns (Transgenomic Application Note 118, http://www.transgenomic.com). Other PCR-related issues remain, which need to be closely examined, such as Hot Start enzymes (for example, AmpliTaq Gold, Perkin-Elmer, Norwalk, CT) that contain antibodies to inactivate the polymerase until the first 95°C step in the thermal cycling run. Hot Start PCR prevents polymerase extension from nonspecific, low-temperature primer binding sites, but artifacts can be introduced into the DHPLC signature due to the additional components (antibodies) in the reaction buffer (M. Nickerson, data not shown).

These examples are mentioned here briefly in the introduction because of the importance of obtaining reproducible signatures from PCR products generated by different labs, which accurately represent the polymorphism of interest. Signature quality in terms of reproducibility and specificity was examined in detail by Nickerson *et al.*[14] but only one high fidelity polymerase was studied. The authors removed DHPLC-related artifacts influencing signatures by adopting a standardized, improved method design, which consisted of addition of a 75 % acetonitrile wash after the analytical gradient, followed by a buffer A (0.1M triethylammonium acetate - TEAA) rinse, and a variable flow rate (Figure 2-1).

This chapter focuses on PCR influences on DHPLC due to choice of polymerase. The choice of polymerase has three variables that can affect DHPLC signature reproducibility. The first, the polymerase reaction buffer, has been discussed above. The other two are polymerase fidelity and PCR annealing temperature. In this chapter, chromatograms were generated from PCR products amplified from SNP carriers and wild type individuals. Chromatograms from

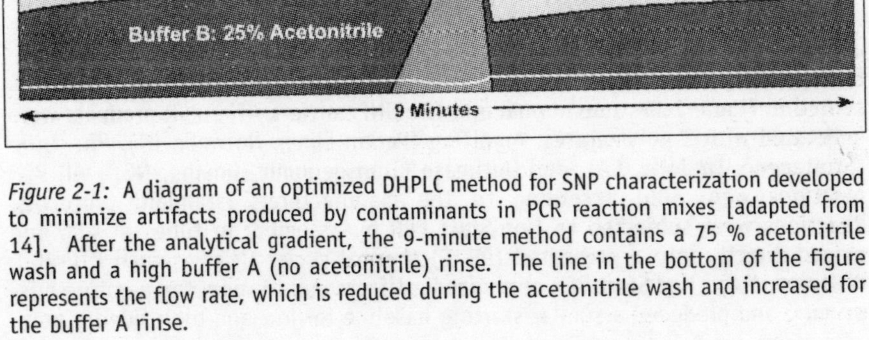

Figure 2-1: A diagram of an optimized DHPLC method for SNP characterization developed to minimize artifacts produced by contaminants in PCR reaction mixes [adapted from 14]. After the analytical gradient, the 9-minute method contains a 75 % acetonitrile wash and a high buffer A (no acetonitrile) rinse. The line in the bottom of the figure represents the flow rate, which is reduced during the acetonitrile wash and increased for the buffer A rinse.

amplicons generated by low fidelity polymerases were compared to elution profiles of products that were amplified using high fidelity enzymes. This allows artifacts in a signature to be directly associated with a polymerase (and its reaction buffer). Additionally, the effect of PCR annealing temperature on DHPLC elution profiles was examined by holding DHPLC conditions constant and varying the temperature of primer annealing during thermal cycling.

Low fidelity polymerases, by definition, exhibit an increased tendency to randomly introduce sequence alterations into amplicons during PCR, termed Random Mutagenesis-PCR (RM-PCR)[15]. Random Mutagenesis-PCR can be useful for DHPLC and other mutation and polymorphism detection techniques, such as SSCP and DGGE, because it provides an effective means of generating positive controls (containing polymorphisms) for examination of instrument performance, method optimization, and to test the accuracy of DHPLC melt algorithm temperature predictions. This has been outlined by Nickerson *et al.*[15] and is presented here with an emphasis on choice of polymerase and PCR conditions to promote efficient synthesis of RM-PCR products.

2.2 Comparison of DNA Elution Profiles Produced by High and Low Fidelity Polymerases

Figure 2-2A shows the pedigree of a family with Birt-Hogg-Dube syndrome (BHD), a disease characterized by aberrant hair follicle development leading to fibrofolliculomas[16]. Recently, the BHD disease phenotype was expanded to include kidney neoplasia[17,18] and an increased risk of lung collapse (spontaneous pneumothorax)[19]. A genome wide scan identified linkage to chromosome 17p11.2[20] and genes in the linked region have been examined for mutations, which cause the disease. In total, 321 amplicons from 39 genes revealed 129 coding SNPs (cSNPs), 49 intronic SNPs, 7 polymorphic repeats (CAG, poly A, etc.), and 6 insertions/deletions[21].

Chromatograms were generated from a heteroduplexed candidate gene amplicon containing a T/G cSNP that was examined for co-segregation with BHD. Data from family members was obtained and three individuals are presented in Figure 2-2A, one normal and two SNP carriers. The PCR products were generated with 3 polymerases, AmpliTaq (Perkin Elmer, Norwalk, CT), Pfu-turbo (Stratagene, La Jolla, CA), and Optimase (Transgenomic, Omaha, NE). All PCR reactions were 50µl according to the manufacturers' standard protocols. Reactions were subjected to Hot Start PCR by assembly in tubes at 4°C and placed directly into a preheated (95°C) thermal cycler (MJ Research PTC-200, Waltham, MA). Hot Start PCR minimized artifacts due to room temperature mispriming and produced a similar starting baseline for low and high fidelity poly-

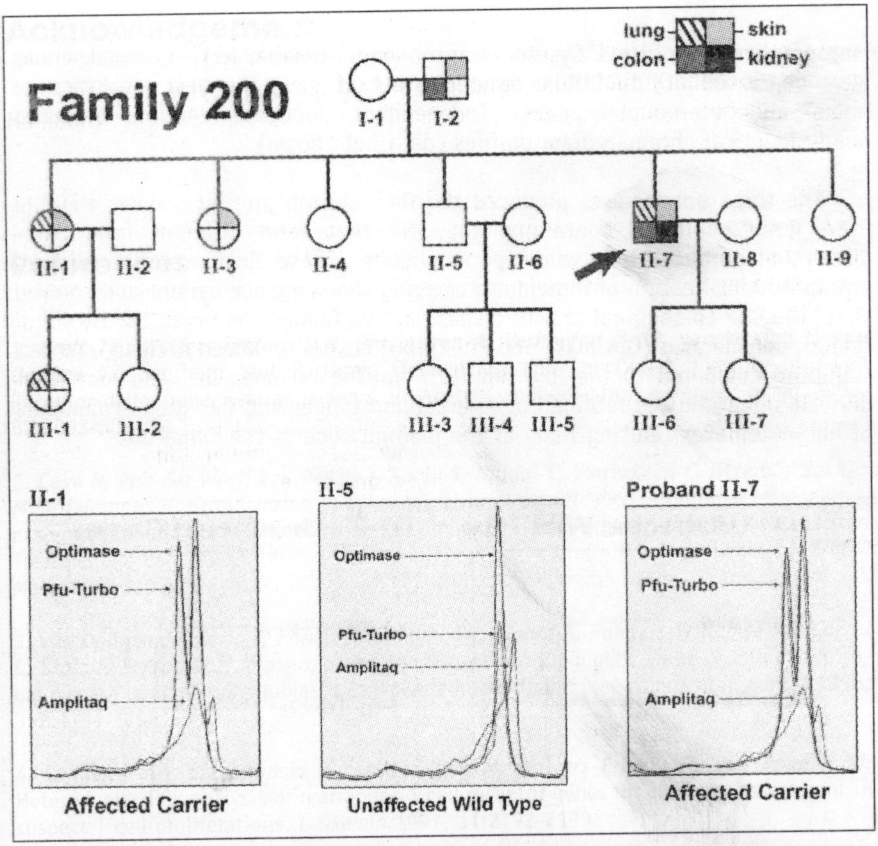

Figure 2-2A: Analysis of SNP co-segregation with BHD disease. A cSNP in a candidate gene was examined by PCR using primers, A1 and A2, and polymerases, AmpliTaq, Optimase, and Pfu-turbo. In the pedigree of Family 200 (top, middle), quadrants refer to aspects of the disease phenotype as shown in the legend at the top (renal tumors, fibrofolliculomas, spontaneous pneumothorax, and colon polyps. The arrow refers to the proband of this family in whom the SNP was first identified. The x-axis is time, in minutes from a 9-minute DHPLC-run, the y-axis is absorbance at 260 nm. The DNA elution profiles that are shown have been extracted from complete chromatograms containing an initial injection spike (appearing at 0.5-1 min) and an acetonitrile clean off peak (at 7 min).

merases. Pfu-turbo, for example, does not require Hot Start due to minimal activity at room temperature. All reactions were amplified at the same annealing temperature for 40 cycles; then 5µl of product was quantitated by 2% agarose gel electrophoresis. The PCR product was heteroduplexed and chromatographed on a DHPLC System (Transgenomic, Omaha, NE). Chromatograms were generated at the DHPLC temperature that gave the best resolution of homo- and heteroduplex peaks. Independent, duplicate reactions produced nearly identical chromatogram profiles (data not shown).

The three polymerases produced the DNA elution profiles shown in Figure 2-2A from amplicons containing the SNP that were different from chromatograms produced from wild type amplicons. These elution profiles allowed accurate identification of individuals carrying the sequence variant and showed that the cSNP co-segregated with disease in this family. However, the two high fidelity polymerases (Optimase and Pfu-turbo) clearly produced a 'clean,' distinct signature compared to the low fidelity AmpliTaq polymerase. The AmpliTaq-derived signature was retained on the column longer and has greater numbers of heteroduplexes (eluting first) at the leading edge of the signature.

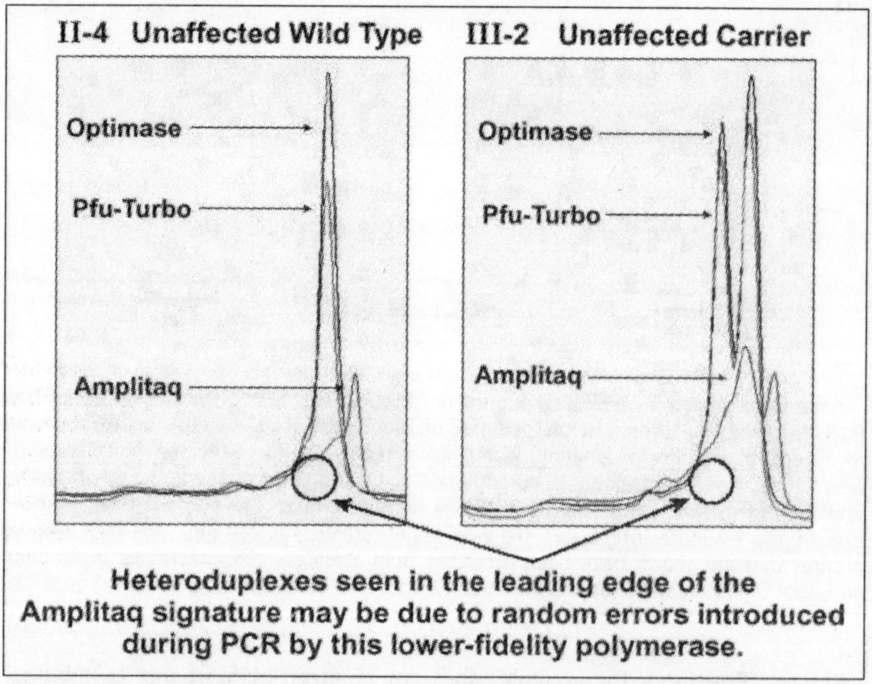

Figure 2-2B: Differences in DNA elution profiles of amplified products from two additional members of Family 200 may be due to polymerase-introduced errors during PCR. The polymorphism co-segregates with disease in this family.

The leading edge heteroduplexes in both the AmpliTaq wild type and SNP carrier chromatograms may be a consequence of an increased population of amplicons containing base substitutions, insertions, and deletions generated by the low fidelity polymerase (Figure 2-5 shows sub-cloned amplicons containing a distribution of nucleotide changes introduced by AmpliTaq). Figure 2-2B focuses on the leading edge heteroduplexes in AmpliTaq-derived chromatograms, which broaden the elution profiles obtained from two additional members of BHD family 200, a wild type and a SNP carrier. It is significant to note the excellent reproducibility of the polymorphic signature and wild type profile between high fidelity polymerases, Optimase and Pfu-turbo. Optimase has an additional advantage because it does not contain components that damage alkylated poly(styrene divinyl-benzene) columns (Transgenomic Application Note 118, www.Transgenomic.com).

2.3 Effects of Polymerase Fidelity on DNA Elution Profiles

Figures 2-2A and 2-2B present minor differences between polymerases that have little effect on DHPLC analysis since the DNA elution profiles are clearly either a wild type, single peak or mutant, double peak. Thus, screening for this polymorphism is not affected by the choice of polymerase when any one of these three enzymes is used.

Figure 2-3 presents DHPLC analyses of an amplicon from either a wild type or heterozygous individual who carries an A/G cSNP. The amplicon spans a gene exon from the BHD critical region and the cSNP was examined for co-segregation with the disease. It produces DHPLC elution profiles that can vary depending on the polymerase used. Amplicons were produced as described above (Hot Start PCR, etc.) except that the PCR annealing temperature was 63°C (Figure 2-3A, 2-3B) or 62°C, 64°C, or 66°C (Figure 2-3C). Figures 2-3A, 2-3B, and 2-3C contain independent, duplicate polymorphic and wild type DNA elution profiles generated using constant DHPLC conditions. Each PCR reaction was quantitated by agarose gel electrophoresis and produced a single band with no background.

Striking differences can be seen (Figure 2-3A and 2-3B) in elution profiles from amplicons produced by individual polymerases. Pfu-turbo, Accutype (Stratagene, La Jolla, CA), and Optimase produce similar wild type and polymorphic elution profiles, while Taq polymerase from Perkin Elmer (AmpliTaq), Qiagen (Valencia, CA) and Invitrogen (Carlsbad, CA) produce PCR products that upon DHPLC exhibit additional peaks in both wild type and SNP carrier samples. The wild type and SNP carrier profiles derived from low fidelity polymerase amplification could all be mistaken for polymorphisms. Arrows highlight additional peaks (compared to the profiles from Figure 2-3B) and high background in the chromatograms.

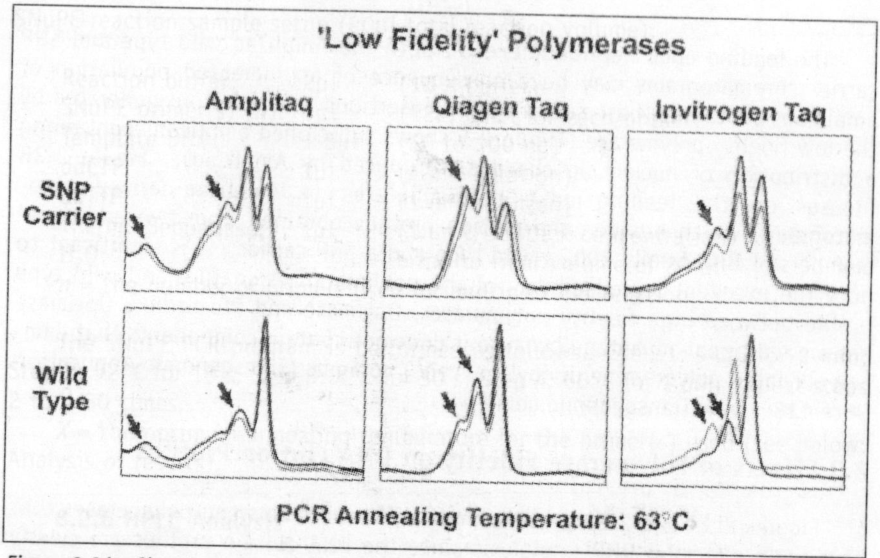

Figure 2-3A: Chromatograms from a heteroduplexed amplicon, FLJ 11+12 (475 bp), containing an A/G cSNP, which was amplified by three low fidelity polymerases at a PCR annealing temperature of 63°C. Chromatograms from two independent PCR reactions are overlaid in each box. Arrows indicate additional peaks or high background. X- and y-axes are defined as in Figure 2A and DNA elution profiles have been similarly extracted from complete chromatograms.

In order to assess the relative contributions of enzyme fidelity and PCR annealing temperature to elution profiles shown in Figure 2-3A, PCR products were amplified using Qiagen Taq and examined by DHPLC to see whether the profile was affected by PCR annealing temperature. It is possible that optimal PCR conditions for amplification of this product (FLJ 11+12) using Taq from Perkin Elmer, Invitrogen, or Qiagen are different from the annealing temperature that is optimal for the high fidelity polymerases (63°C). The PCR products in Figure 2-3C were amplified using Qiagen Taq at the PCR annealing temperature shown below the boxed SNP carrier and wild type DHPLC elution profiles. As shown, artifacts in the chromatogram were not reduced or removed even though the PCR annealing temperature was raised to 66°C and a single band with no background was seen by agarose gel electrophoresis.

Why does the elution profile vary? Apparently, all or some combination of the following may play interacting roles:

1. Enzyme storage buffer
2. PCR reaction buffer

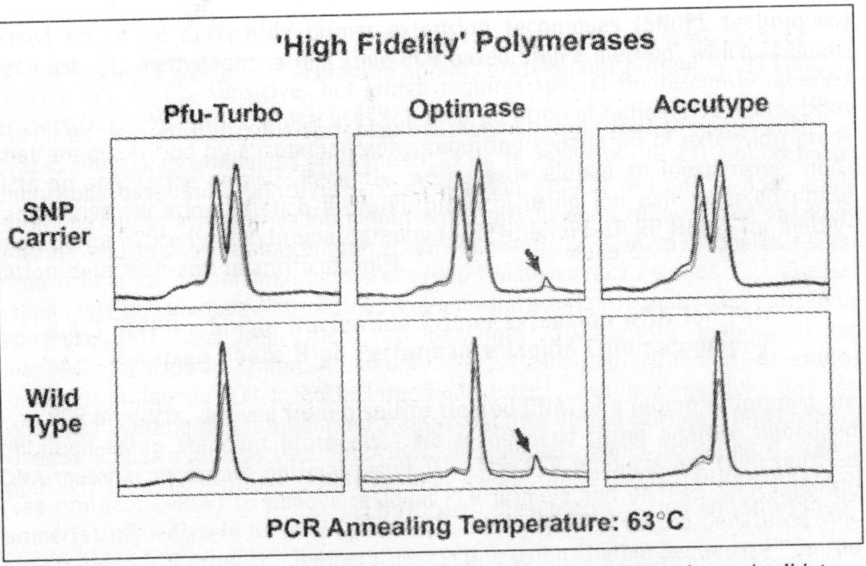

Figure 2-3B: PCR with high fidelity polymerases produced SNP carrier and wild type DHPLC elution profiles that were different from profiles generated from PCR products amplified by low fidelity polymerases. PCR annealing temperature is constant at 63°C.

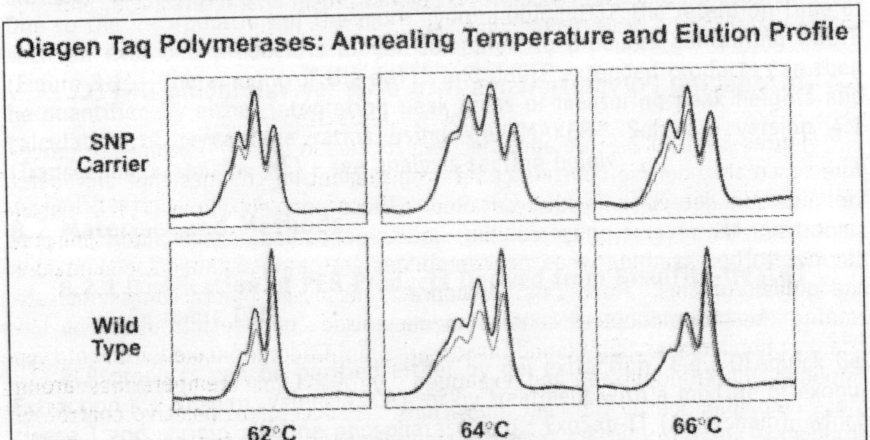

Figure 2-3C: DHPLC elution profiles from heteroduplexed Qiagen Taq amplified PCR products at PCR annealing temperatures 62°, 64°, and 66°C.

3. Polymerase fidelity (Figure 2-3A and 2-3B)
4. PCR annealing temperature (minor influence, if any, see Figure 2-3C).

It is apparent that large numbers of nucleotide substitutions may be introduced by low fidelity enzymes at all PCR annealing temperatures (Figure 2-3C). Since each of the low fidelity polymerases in Figure 2-3A produced additional peaks at 63°C, while none of the three high fidelity enzymes produced these same artifacts (Figure 2-3B, Optimase has a unique, minor peak of late eluting material), it appears that enzyme fidelity is a dominant factor. It is also possible that secondary structure that may exist in the DNA sequence of the amplicon is handled differently by high and low fidelity enzymes. This may contribute to the additional peaks seen in Figures 3A and 3C in both the SNP carrier and wild type samples. Lastly, the contribution of PCR annealing temperature to DHPLC elution profile artifacts cannot be ruled out completely. Taq polymerase from Perkin Elmer, Qiagen, and Invitrogen may optimize at PCR annealing temperatures above those shown here so that a profile matching the high fidelity enzyme panel is produced upon DHPLC.

To conclude this section, the type of polymerase with the manufacturer's reaction buffer, and possibly an enzyme-specific PCR annealing temperature, as well as DHPLC temperature and acetonitrile gradient are now the critical parameters for accurate data reproducibility, polymorphism characterization, and signature matching by scatter plot analysis.

2.4 Introducing Polymorphisms into DNA for Validating DHPLC

Figures 2-4 and 2-5 are adapted from Nickerson *et al.*[15] where the procedure to create positive controls containing nucleotide changes that are useful for mutation detection is fully detailed. These controls allow DHPLC instrumentation, the reverse phase column, and the accuracy of melt algorithm predictions to be examined when new amplicons are being scanned for mutations and polymorphisms. Here, this procedure is discussed in the context of subcloning the heteroduplexes containing nucleotide changes introduced by low fidelity polymerases. Sub-cloned inserts are amplified and mixed with wild type amplicon, heteroduplexed, and examined by DHPLC at temperatures around those recommended by the melt algorithms. A successful positive control will produce a signature distinct from the single-peaked wild type profile, and distinct from polymorphic signatures of sub-clones containing nucleotide substitutions located in other parts of the amplicon. The exact location of polymorphisms is determined by sequencing, but in nine minutes per clone the DHPLC will provide detailed information about whether a polymorphism is present.

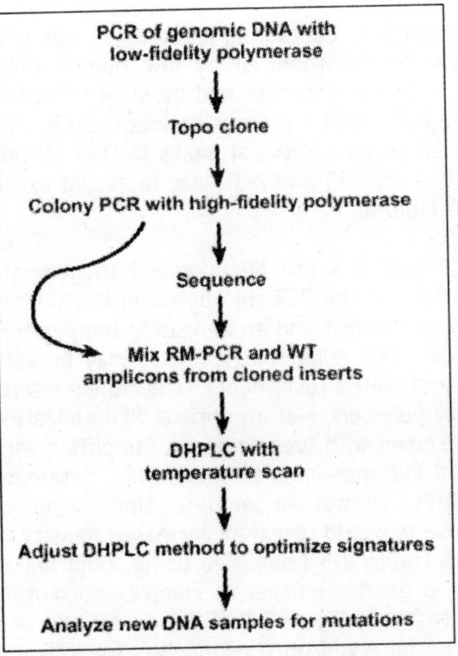

Figure 2-4: The procedure for creating amplicons containing polymerase-introduced, random polymorphisms in a DNA sequence of interest [adapted from 15]. TOPO clone refers to topoisomerase mediated (5 min) ligation of an amplicon into a prepared cloning vector (Invitrogen). This procedure utilizes low fidelity PCR conditions to generate positive controls that validate DHPLC instrumentation, column integrity, method design, and melt algorithm predictions.

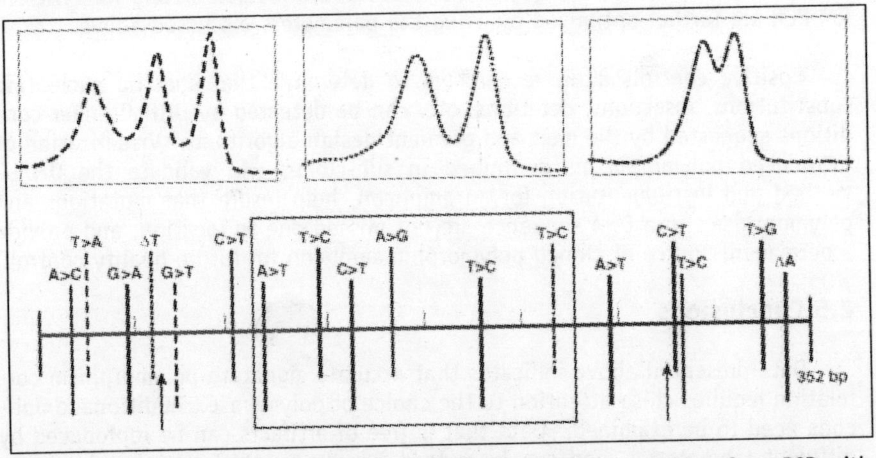

Figure 2-5: Locations of introduced polymorphisms (bottom) resulting from PCR with AmpliTaq of a 352 bp product (MET exon 15) [adapted from 15]. Unique signatures from three cloned inserts amplified with a high fidelity polymerase (Pfu-turbo) are shown (above). Polymorphic signatures and the locations of the corresponding nucleotide substitutions are shown in matching dashed and dotted lines.

Figure 2-3 suggests that PCR reactions, containing molecules with mismatches and substitutions that have been introduced by low fidelity polymerases, can be visualized by DHPLC. This is accomplished by identifying fast eluting, additional peaks in chromatograms that represent heteroduplexes containing mismatches. Amplicons visualized as additional peaks in DNA elution profiles from Figures 2-3A (marked with arrows) and 2-3C may represent excellent fractions to be collected for sub-cloning.

Thus, the protocol outlined in Figure 2-4 can be improved to generate greater numbers of nucleotide alterations at the PCR step by using low fidelity enzymes to amplify the DNA segment of interest and an annealing temperature that is several degrees below optimal. PCR reactions (2 hours) may be subcloned using topoisomerase (1 hour) and plated (overnight). The cloned inserts are then amplified with a high fidelity polymerase at an optimal PCR annealing temperature (2 hours), mixed with a known wild type amplicon (amplified with high fidelity polymerase at an optimal PCR annealing temperature!), heteroduplexed (30 min), and examined by DHPLC (9 min per sample). Upon sequencing, it is expected that the cloned inserts would reveal an increased density of nucleotide changes over that seen in Figure 2-5 (amplified using AmpliTaq at an optimal temperature). Hopefully, a greater number of complex nucleotide changes (such as delAGinsC, see Family 200 in Figure 2-2 of[21]), insertions, and deletions (in addition to SNPs) will be observed once parameters for efficient RM-PCR are further optimized by using the parameters discussed above.

Positive controls allow researchers to determine that specific nucleotide substitutions, insertions, deletions, etc. can be detected by DHPLC under conditions suggested by the melt and gradient design algorithms. Visualization of introduced polymorphisms contained in sub-clones will validate the DHPLC method and instrumentation for an amplicon, help insure that mutations and polymorphisms in a DNA sequence are not missed due to location, and provide a permanent source of cloned polymorphic amplicon for future quality control.

2.5 Conclusions

Data presented above indicates that accurate signature-polymorphism correlation requires close attention to the choice of polymerase. Additional amplicons need to be examined. Data that is free of artifacts can be reproduced by different laboratories, and can be judged equally on platforms that integrate mutation detection data from different assays.

Scatter plot analysis of large numbers of chromatograms will be necessary for patient mutation detection, high throughput genotyping, population studies of polymorphic markers, epidemiology, and pharmacogenomics. DHPLC scatter plot analysis is only accurate if the DNA elution profiles can be reproduced. The data presented here shows that wild type amplicons can be easily distinguished from SNP carriers but also suggests that some amplicons may require identification of an enzyme-specific signature that corresponds to a nucleotide alteration. The fact that a polymorphic signature can change in response to different PCR polymerases indicates that this parameter must be discussed in methods describing mutation detection by DHPLC.

PCR annealing temperature was not mentioned in a recent review[1] and a survey of recent DHPLC publications indicated many instances where this critical information was lacking[22-25]. Many researchers have chosen to go with touchdown procedures where the PCR annealing temperature decreases by a specific amount every 1-5 cycles during amplification [for example[26-28] and this approach may suitably address any effects that annealing temperature may have on elution profiles. The DHPLC community assumes that all elution profiles presented in a paper are the result of optimized PCR reactions. Additional experiments may show that artifacts resulting from poorly optimized PCR reactions may lead to inaccurate polymorphism characterization by DHPLC.

In most cases there is little reason to switch polymerases for an amplicon, which has already been successfully amplified. However, false positives are sometimes seen in DHPLC when screening amplicons and the data presented above contributes to understanding why some PCR products consistently show artifacts in wild type samples. This is likely to be mainly the result of errors introduced by low fidelity polymerases during PCR.

The importance of the effort to create validated amplicons for every SNP, gene exon, and interesting non-coding genomic region cannot be understated (for example, see Applied Biosystem's web page < www.allsnps.com >). However, this project must expand to the public domain, in addition to the private sector, and be generalized to include information relevant to various types of SNP detection assays, such as DHPLC, sequencing, primer extension, nuclease cleavage, etc. SNP discovery[29] is not enough to insure effective use of SNPs for understanding disease[30], if the detection method(s) are not reliable. Since most assays are PCR-based, researchers need to determine and describe optimal amplification conditions, including primer sequences, polymerase, annealing temperature, and validated assay parameters in order to ensure reproducibility. A database that attempts to satisfy the above listed criteria is freely available at http://www.mutationdiscovery.com. Researchers need to submit this information to the various human, mouse, drosophila, etc. (public and private) genome databases so that it is freely accessible and readily available.

After this chapter was submitted (August 2002) a related paper was published in December 2002: Muhr D, Wagner T, Oefner PJ: Polymerase chain reaction fidelity and denaturing high-performance liquid chromatography. J. Chromatogr. B 2002;782:105-110.

Acknowledgements

Thanks to BHD gene team members. Dr. Laura Schmidt critically reviewed the manuscript and designed primers using Oligo Tech ver. 1 (Oligos Etc & Oligo Therapeutics), which were synthesized by Operon Technologies (Alameda, CA). Michelle Warren, M.S., optimized PCR conditions, analyzed Big Dye Terminator sequencing (ABI, Foster City, CA) for SNP detection using Lasergene (DNAStar, Madison, WI), and characterized TOPO sub-clones (Figure 4, Invitrogen, Carlsbad, CA). Dr. Mike Bedford and Ms. Laura Bunnell provided technical assistance for Transgenomic's WAVE(r) Nucleic Acid Fragment Analysis System (Omaha, NE) and excellent discussion of mutation detection. Thanks most of all to Laboratory of Immunobiology Chief, Dr. Bert Zbar for critically reviewing this manuscript and for providing an excellent research environment at the National Cancer Institute-Frederick

Disclaimer

This research was supported by the National Cancer Institute (NCI) and includes a list of reagents employed in the research plan. No endorsement of these reagents by the federal government is intended nor should be inferred.

References

1. ***Xiao W, Oefner PJ:*** Denaturing high-performance liquid chromatography. *Hum Mutat 2001*;17:439-474.

2. ***Ahmadian A, Lundeberg J:*** A brief history of genetic variation analysis. *BioTechniques 2002*;32:1122-1137.

3. ***Kristensen VN, Kelefiotis D, Kristensen T, Borresen-Dale A-L:*** High-throughput methods for detection of genetic variation. *BioTechniques 2001*;30:318-332.

4. ***Hamilton SC, Farchaus JW, Davis MC:*** DNA polymerases as engines for biotechnology. *BioTechniques 2001*;31:370-383.

5. ***Cline J, Braman JC, Hogrefe HH:*** PCR fidelity of Pfu DNA polymerase and other thermostable DNA polymerases. *NAR 1996*;24:3546-3551.

6. ***Seaton B:*** Some like it hot: thermophilic DNA polymerases. *BioConsumer Review 1996*;Spring:48-51.

7. ***Ravnik-Glavac M, Atkinson A, Glavac D, Dean M:*** DHPLC screening of cystic fibrosis gene mutations. *Hum Mutat 2002*;19:374-383.

8. ***Ezzeldin H, Okamoto Y, Johnson MR, Diasio RB:*** A high-throughput DHPLC method for the identification of variant alleles associated with dihydropyrimidine dehydrogenase deficiency. *Anal Biochem 2002*;306:63-73.

9. ***Baumer A, Wiedemann U, Hergersberg M, Schinzel A:*** A novel MSP/DHPLC method for the investigation of the methylation status of imprinted genes enables the molecular detection of low cell mosaicisms. *Hum Mutat 2001*;17:423-430.

10. ***Keller G, Hartmann A, Mueller J, Hofler H:*** DHPLC for the analysis of somatic p53 mutations. *Lab Invest 2001*;81:1735-1737.

11. ***Benit P, Bonnefont J-P, Mostefa AK, Francannet C, Munnich A, Ray PF:*** DHPLC-based prenatal diagnosis for tuberous sclerosis. *Prenat Diag 2001*;21:279-283.

12. ***Liu MR, Pan KF, Li ZF, Wang Y, Deng DJ, Zhang L, Lu YY:*** Rapid screening of mitochondrial DNA mutations by using DHPLC. *World J Gastroenterol 2002*;8:426-430.

13. ***Walter J, Lamb H, Griffiths C, Hecker K, Robinson M, Daniels M, Krause P, Gjerde D, Taylor P:*** Evaluation of PCR enzyme related parameters affecting DHPLC. Poster presented at the 2001 British Society for Human Genetics meeting.

14. ***Nickerson ML, Weirich G, Zbar B, Schmidt LS:*** Signature-based analysis of MET proto-oncogene mutations using DHPLC. *Hum Mutat 2000*;16:68-76.

15. ***Nickerson ML, Warren MB, Zbar B, Schmidt LS:*** Random mutagenesis-PCR to introduce alterations into defined DNA sequences for validation of SNP and mutation detection methods. *Hum Mutat 2001*;17:210-219.

16. ***Birt AR, Hogg GR, Dube' WJ:*** Hereditary multiple fibrofolliculomas with trichodiscomas and acrochordons. *Arch Dermatol 1977*;113:1674-1677.

17. ***Toro JR, Glenn G, Duray P, Darling T, Weirich G, Zbar B, Linehan WM, Turner ML:*** Birt-Hogg-Dube syndrome: a novel marker of kidney neoplasia. *Arch Dermatol 1999*; 135:1195-1202.

18. ***Lindor NM, Hand J, Burch PA, Gibson LE:*** Birt-Hogg-Dube syndrome: an autosomal dominant disorder with predisposition to cancers of the kidney, fibrofolliculomas, and focal cutaneous mucinosis. *Int J Dermatol 2001*;40:653-656.

19. ***Zbar B, Alvord WG, Glenn G, Turner M, Pavlovich CP, Schmidt L, Walther M, Choyke P, Weirich G, Hewitt SM, Duray P, Gabril F, Greenberg C, Merino MJ, Toro J, Linehan WM:*** Risk of renal and colonic neoplasms and spontaneous pneomothorax in the Birt-Hogg-Dube syndrome. *Cancer Epidemiol Biomarkers Prev 2002*;11:393-400.

20. ***Schmidt LS, Warren MB, Nickerson ML, Weirich G, Matrosova V, Toro JR, Turner ML, Duray P, Merino M, Hewitt S, Pavlovich CP, Glenn G, Greenberg CR, Linehan WM, Zbar B:*** Birt-Hogg-Dube syndrome, a genodermatosis associated with spontaneous pneumothorax and kidney neoplasia, maps to chromosome 17p11.2. *Am J Hum Genet 2001*;69:876-882.

21. ***Nickerson ML, Warren MB, Toro JR, Matrosova V, Glenn G, Turner ML, Duray P, Merino M, Choyke P, Pavlovich CP, Sharma N, Walther M, Munroe D, Hill R, Maher E, Greenberg C, Lerman MI, Linehan WM, Zbar B, Schmidt LS:*** Mutations in a novel gene lead to kidney tumors, lung wall defects, and benign tumors of the hair follicle in patients with the Birt-Hogg-Dube syndrome. *Cancer Cell 2002*;2:157-164.

22. ***Boutin P, Vasseur F, Samson C, Wahl C, Froguel P:*** Routine mutation screening of HNF-1 and GCK genes in MODY diagnosis: How effective are the techniques of DHPLC and direct sequencing used in combination? *Diabetologia 2001*;44:775-778.

23. ***Sauer CG, White K, Stohr H, Grimm T, Hutchinson A, Bernstein PS, Lewis RA, Simonelli F, Pauleikhoff D, Allikmets R, Weber BHF:*** Evaluation of the G protein coupled receptor-75 (GPR75) in age related macular degeneration. *Br J Ophthalmol 2001*;85:969-975.

24. ***Taliani MR, Roberts SC, Dukek BA, Pruthi RK, Nichols WL, Heit JA:*** Sensitivity and specificity of DHPLC for unknown protein C gene mutations. *Genet Test 2001*;5:39-44.

25. ***Hecker KH, Asea A, Kobayashi K, Green S, Tang D, Calderwood SK:*** Mutation detection in the human HSP70B' gene by DHPLC. *Cell Stress Chaperones 2000*;5:415-424.

26. ***Thorstenson YR, Shen P, Tusher VG, Wayne TL, Davis RW, Chu G, Oefner PJ:*** Global analysis of ATM polymorphism reveals significant functional constraint. *Am J Hum Genet 2001*;69:396-412.

27. ***LeGac G, Mura C, Ferec C:*** Complete scanning of the hereditary hemochromatosis gene (HFE) by use of DHPLC. *Clin Chem 2001*;47:1633-1640.

28. ***Narayanaswami G, Taylor PD:*** Improved efficiency of mutation detection by DHPLC using modified primers and hybridization procedure. *Genet Test 2001*;5:9-16.

29. ***Weiner MP, Hudson TJ:*** Introduction to SNPs: discovery of markers for disease. *Biotechniques 2002*; Jun suppl:4-13.

30. ***Schifreen RS, Storts DR, Buller AM:*** The challenge of using SNPs in the understanding and treatment of disease. *Biotechniques 2002*; Jun suppl:14-21.

SECTION II

GENE SPECIFIC APPLICATIONS OF MUTATION DETECTION BY DHPLC

CHAPTER

3

DENATURING HIGH PERFORMANCE LIQUID CHROMATOGRAPHY (DHPLC), AN IMPORTANT MUTATION DETECTION TECHNIQUE IN MUTATION ANALYSIS OF THE BREAST AND OVARIAN CANCER PREDISPOSING GENES BRCA1 AND BRCA2

Norbert Arnold[1] and Dieter Niederacher[2]

Departments of Gynecology and Obstetrics, University Hospitals Kiel[1] and Düsseldorf[2], Germany

Contact:
Prof. Dr. rer. nat. Norbert Arnold
Universitätsklinikum Kiel, Klinik für Gynaekologie und Geburtshilfe,
Onkologisches Labor, Michaelisstraße 16 , 24105 Kiel, Germany
E-mail: nkarnold@email.uni-kiel.de

0-9664027-7-4/03/$0.00+$.50 *From:* **Genetic Variance Detection~Nuts & Bolts of DHPLC in Genomics** (pp.53-78)
Edited by: Karl H. Hecker, Ph.D.

3.1 The BRCA Genes and Hereditary Breast and Ovarian Cancer

During the last decade several genes responsible for autosomal dominant transmission of greatly increased risk of specific cancers were identified. Although it has been known for years that some families exhibited an apparently autosomal dominant pattern of early-onset breast and ovarian cancer, the characterization of the syndrome of hereditary breast and ovarian cancer (HBOC) predominantly resulted from the identification of the BRCA1 and BRCA2 genes[1,2] and the ability to analyze the BRCA genes for mutations responsible for the HBOC syndrome. Mutations in both genes are highly penetrant and confer an increased risk not only for HBOC, but also for other malignancies, e.g. colon and prostate cancer. Early comprehensive studies of the Breast Cancer Linkage Consortium assessed the contribution of BRCA1 and BRCA2 to inherited breast cancer by linkage and mutation analysis in families with at least 4 cases of breast cancer.[3] In an estimated 52% of families disease was linked to BRCA1, in 32% of families to BRCA2, and to neither gene in 16% indicating other predisposing genes as well. In the majority of families (81%) with both cancers, breast and ovarian, cancer was linked to BRCA1. In most other cases cancer was linked to BRCA2. The majority of families (76%) with male breast cancer cases were caused by mutations in the BRCA2 gene. The largest proportion of BRCA1/BRCA2 negative families (67%) were families with female breast cancer only. Testing large series of families in a clinical setting, e.g. the large extended families in which the BRCA genes were originally discovered, with multiple individuals who had developed breast or ovarian cancer at an early age, were found not to be typical of most BRCA families[4]. A recent study aimed at assessing characteristics of family history that correlate best with the presence of BRCA1 and BRCA2 mutations in individuals tested in a clinical setting[5]. Mutations were identified in 17.2 % of the 10,000 individuals tested, including 968 (20%) of 4,843 women with breast cancer and 281 (34%) of 824 women with ovarian cancer. Mutation prevalence in high-risk women of European ancestry was 16% (712/4379). This is comparable with the risk of women of other, non-Ashkenazi ancestries. Interestingly, 36% of individuals tested in this study reported a family history of only a single first-degree or second-degree relative with early-onset breast or ovarian cancer. Among the latter, 13% (224/1,693) carried deleterious mutations indicating that family history of BRCA1 or BRCA2 mutation carriers ascertained in a clinical setting with a certain amount of small family pedigrees may be neither dramatic nor obvious.

An essential prerequisite for achieving clinical benefits from the cloning of the BRCA genes is finding and assessing the significance of mutations in these susceptibility genes. In order to coordinate the detection and interpretation of BRCA mutations and to make the results available to as many qualified investigators as possible, a central repository of information regarding mutations and polymorphisms, the Breast Cancer Information Core (BIC), was created and is being maintained by an international collaborative effort hosted by NHGRI (http://research.nhgri.nih.gov/bic/). A total of 6672 and 5624 sequence variations/entries for BRCA1 and BRCA2, respectively, have been registered with the BIC database. These entries represent 1237 and 1381 distinct mutations, polymorphisms and variants of the BRCA1 and BRCA2 gene, respectively. A total of 711 and 872 sequence variants for BRCA1 and BRCA2, respectively, were reported only once. Most of the distinct alterations reported are deleterious mutations such as frame shift, nonsense (574 for BRCA1 and 576 for BRCA2), or splice site mutations (132 for BRCA1 and 69 for BRCA2).

Implementation of informative BRCA1 and BRCA2 testing programs is aided by the acquisition of population-specific genetic data. In Europe, such a comprehensive study with more than 400 individuals has been performed only for the BRCA1 gene in the Dutch population.

In other populations like the French, Swedish and the Finnish, 100-200 families have been screened for aberrations in the BRCA1 gene. In each of these populations less than 150 families have been investigated for mutations in the BRCA2 gene. The proportion of cases attributed to BRCA1 ranges from 10% in Finland and the Netherlands to approximately 25% in France and Sweden. BRCA2 mutation detection rates are generally around 10%, with the exception of Italy where germline mutations have been found in as much as 25% of 49 families analyzed. In the German population, where most studies have been limited to the analysis of BRCA1 or to relatively small series, a comprehensive study was initiated in 1996 by the formation of a research collaboration, the "German Consortium for Hereditary Breast and Ovarian Cancer," supported by the Deutsche Krebshilfe (German Cancer Aid).

3.2 Organization and Aims of the German Consortium

With the possibility of testing for alterations in the two BRCA genes known to predispose for breast and ovarian cancer, identification of individuals at risk and their inclusion in surveillance programs became feasible. Careful evaluation of paternal as well as maternal family history is required, especially in women diagnosed with breast cancer before the age of 50 or ovarian cancer at any age, to enable the appropriate identification and counseling of individuals at risk for carrying mutations in BRCA1 and BRCA2. Breast and/or ovarian cancer patients who belong to families with multiple cases of early onset breast and ovarian cancer are strong candidates for BRCA mutations.

BRCA testing is predictive because a given BRCA mutation does not inevitably lead to cancer in a given individual. A BRCA mutation markedly increases the likelihood of developing breast and ovarian cancer, but other genetic factors and environmental influences also contribute to the phenotype. A BRCA mutation predicts a higher lifetime risk for cancer, but does not make the diagnosis a certainty. Similarly, a true negative result, e.g. the absence of a detectable mutation, for a member of a high-risk family predicts that the risk of cancer is not markedly increased over that of the average woman in the population. Therefore, the latter still have a probability of 10 to 12% of developing sporadic breast cancer and 1-2% of sporadic ovarian cancer.

Testing patients at high risk in an established clinical trial requires standardized procedures and confidentiality. The impact of breast and ovarian cancers on patients and their families can be considerable. Exploring a patient's emotional and psychological concerns before initiating tests is recommended. In addition, genetic counseling is not just for the individual being tested, but is directed at the entire family. In extending consultation to interested family members the possible ethical-philosophical implications (informed consent, personal autonomy), psychosocial implications (perception of risk, emotional state), and legal implications must be kept in mind[6]. Therefore, the complexities surrounding BRCA testing suggested an interdisciplinary counseling strategy.

Taking the aforementioned considerations into account, the German Cancer Aid decided to fund ten research centers in Germany starting in 1997. To ensure ease of access for families, centers were distributed all over Germany - Berlin, Bonn, Düsseldorf, Frankfurt, Kiel, Heidelberg, München, Münster, Ulm, and Würzburg. In the following years two additional centers in the eastern part of Germany - Dresden and Leipzig - were included. All centers were committed to the same inclusion criteria and counseling strategies. The study protocol uses a multidisciplinary approach that provides counseling and genetic testing by bringing together, in each center, a team of genetic counselors, gynecologists and psycho-oncologists. Each local ethics committee approved the protocol used. Bilateral cancer or breast and ovarian cancer of the same patient were counted as independent cases of breast/ovarian cancer. Inclusion criteria were based on family history as follows: Families with 2 or more cases of breast cancer, which included at least one case with onset under the age of 50 years; families with one or more cases of breast and at least 1 case of ovarian cancer; families with at least 1 male breast cancer, and families with a single case of breast cancer with age of diagnosis before 35 years. Representatives of the three disciplines performed counseling in either three consecutive sessions or in one session with all members present. Referrals to the study have come from university hospitals, private practice physicians, as well as by self-referral. After a first counseling session, eligible candidates had a minimum of four weeks to consider participating in the study. After a second counseling session, blood was drawn for genetic testing. Test results were presented in a third counseling session.

In the beginning different strategies for the analysis of the BRCA genes were used in the individual centers. Some centers uses SSCP and/or PTT as pre-screening methods followed by DNA-sequencing of positive results. Others routinely performed direct DNA-sequencing. At the beginning the center in Kiel, in collaboration with Peter Oefner at Stanford University, tested the usefulness of the DHPLC technique as a prescreening method for BRCA analysis. After initial, promising results, the center in Kiel initiated a comparative study between SSCP, DHPLC, and sequencing. The results[7,8] were presented to the other centers and convinced 10 of them to change their analysis strategy according to the Kiel procedure.

3.3 Established Techniques in (BRCA1/2) Mutation Analysis

Mutation detection has found important applications in functional genomics and clinical diagnostics. Realizing the full benefits of the decoding of the human genome requires rapid and sensitive methods for screening genomic DNA for unknown sequence variations. Technologies used for the latter task should be capable of fully automated, high-throughput analysis that ideally does not require modified PCR primers, customized specific reagent arrays, detection labels, or any sample pre-treatment other than PCR. Numerous techniques are available for mutation detection like direct sequencing (DS), protein truncation test (PTT)[9], single-strand conformation polymorphism (SSCP)[10], dideoxy fingerprinting assay (DDF)[11], denaturing gradient gel electrophoresis (DGGE)[12], two-dimensional gene scanning (TDGS)[13], conformation-sensitive gel electrophoresis (CSGE)[14], enzymatic mutation detection (EMD)[15], allele-specific oligonucleotide hybridization (ASO)[16], and immobilized DNA hybridization assays[17]. Up to now, none of these methods meet the challenges set forth above. The widely used SSCP analysis is economical and simple, but has low sensitivity, ranging from 60 to 90%. However, the major disadvantage of SSCP is the lack of its objective predictability. Thus, since there are no theoretical assurances, non-appearance of a band-shift does not prove the absence of a mutation. By contrast, automated direct sequencing is highly sensitive but costly and labor intensive. DDF is a combination of a Sanger sequencing reaction with multiple-fragment SSCP and is more sensitive than SSCP alone, but still labor intensive. With PTT only sequence alterations leading to a truncated protein can be detected. Other alterations (missense, inframe deletions, and insertions) escape the detection of the PTT assay. Other methods, including DGGE and CSGE, have been used with some success, but despite their improved sensitivity over SSCP, these methods are technically challenging and formatted for manual use only. TDGS is a method for analyzing multiple DNA fragments in parallel for all possible sequence variations, using extensive multiplex PCR and two-dimensional electrophoretic separation on the basis of size and melting temperature. One source of error is the interpretation of the complex spot patterns produced by this method. Also, the sensitivity of TDGS is impaired by frequent preferential amplification of the non-mutant allele, resulting in very "light" heteroduplices that obscured accurate reading of the gels[18]. EMD methods for mutation scanning still lack the sensitivity and specificity of the chemical cleavage of the mismatch method. Other approaches such as immobilized DNA hybridization arrays still have significant false positive signal, as well as high cost per assay.

Denaturing high performance liquid chromatography (DHPLC) has generated increased interest over the past years in clinical genetics, because of its potential for automation and its ease of application. The technique, which is based on heteroduplex detection, allows for automated identification of single-nucleotide substitutions and small deletions or insertions. Heteroduplex profiles are easily distinguished from homoduplex peaks[19]. Furthermore, DHPLC has been shown to clearly resolve mutations in various genes with detection rates ranging from 92.5 to 100%[20]. An updated version of the list of genes screened entirely or partly by DHPLC is available at the web site of the Stanford Genome Technology Center, Palo Alto, USA (http://insertion.stanford.edu/dhplc_genes1.html).

3.4 Development of a DHPLC Based BRCA1/2 Gene Test in Routine Genetic Diagnostics

The two BRCA genes together comprise approximately 15,700 nucleotides of open reading frame. More than 1,000 mutations of deduced or established clinical significance have been identified and are distributed throughout the 48 coding exons and respective splice junctions of the two genes. In families with both breast and ovarian cancer the detection rate for BRCA1 and BRCA2 mutations is only 43% and 10%, respectively[21]. The BRCA mutation frequencies in most families were even less because of the inclusion criteria for BC/OC families to be tested within the German study, which required one case of early-onset breast cancer (< 50 years) and one first relative with breast cancer at any age. Therefore, a selective prescreening to avoid labor and cost intensive sequencing of wild type exons is an indispensable necessity.

3.5 Number of Amplified BRCA1/BRCA2 Fragments

Primer pairs for the amplification of BRCA1 and BRCA2 exons and intron-boundary regions were chosen with respect to the length and optimal DHPLC performance of the PCR products. Samples suitable for mutation detection by

DHPLC are PCR products in the range 200 - 500 base pairs (bp). Fragments larger than 500 bp can be run, but sensitivity for mutation detection may be reduced[22]. Fragments of less than about 200 bp in length do not give satisfactory results, as the fragments melt over too narrow a temperature range.

For oligonucleotide design, software provided through the world wide web at http://www.genome.wi.mit.edu/genome_software/other/primer3.html or at http://genome-www2.stanford.edu/cgi-bin/SGD/web-primer were utilized. The underlying sequences were based on GenBank entries for BRCA1 (L78833) and BRCA2 (Z74739). As shown in ***Table 3-1*** and ***Table 3-2***, the BRCA1 gene was divided into 34 PCR fragments. Lengths of the fragments ranged from 234 to 564 bp. The lengths of the 42 PCR fragments dividing the BRCA2 gene ranged from 278 to 707 bp. Some of the oligonucleotide primers were published earlier[7,8]. The primer pairs described are those used by the majority of the centers of the German consortium. PCR needs to be optimized, because poor quality PCR products will produce poor quality mutation detection/DHPLC results.

3.6 Establishing Running Conditions

Separation of homoduplex and heteroduplex DNA, and hence mutation detection, by DHPLC is temperature dependent. Selecting appropriate temperatures for the analysis of each individual fragment is critical to the success of DHPLC. There are three different methods for predicting analysis temperatures. One possible method is the generation of an experimental melting curve for determining optimal analysis temperatures. In order to generate an experimental melting curve, a wild type fragment is run at progressively increasing temperatures as shown in Figure 3-1A. The temperature steps (1°C) in the example given here for BRCA2, exon 4 cover a temperature range from 50 to 56°C. For the example presented here, a significant reduction in retention time is observed at each temperature step. At temperatures above 56°C no further effect on retention time is observed under the gradient conditions employed here, because the DNA fragment analyzed is completely denatured. On the basis of this data set, a melting curve can be constructed by plotting retention time as a function of analysis temperature with end points at 50°C, corresponding to 100 % double-strand character of the fragment analyzed and 56°C, corresponding to 100 % single-strand character of the fragment analyzed. In this particular case, the temperature point at which 50 % of the fragment analyzed was denatured was determined to be 53°C. The analysis temperature is generally set one to two degree below the experimentally determined temperature point at which 50 % of the fragment is denatured. Figure 3-1B shows the elution profiles of a positive control (BRCA2, IVS4+67A->C) and the detection range for this mutation (51-55°C). The "best" result, most clearly distinguishing/resolving both hetero- and homoduplices is seen at 53°C. This experimental approach

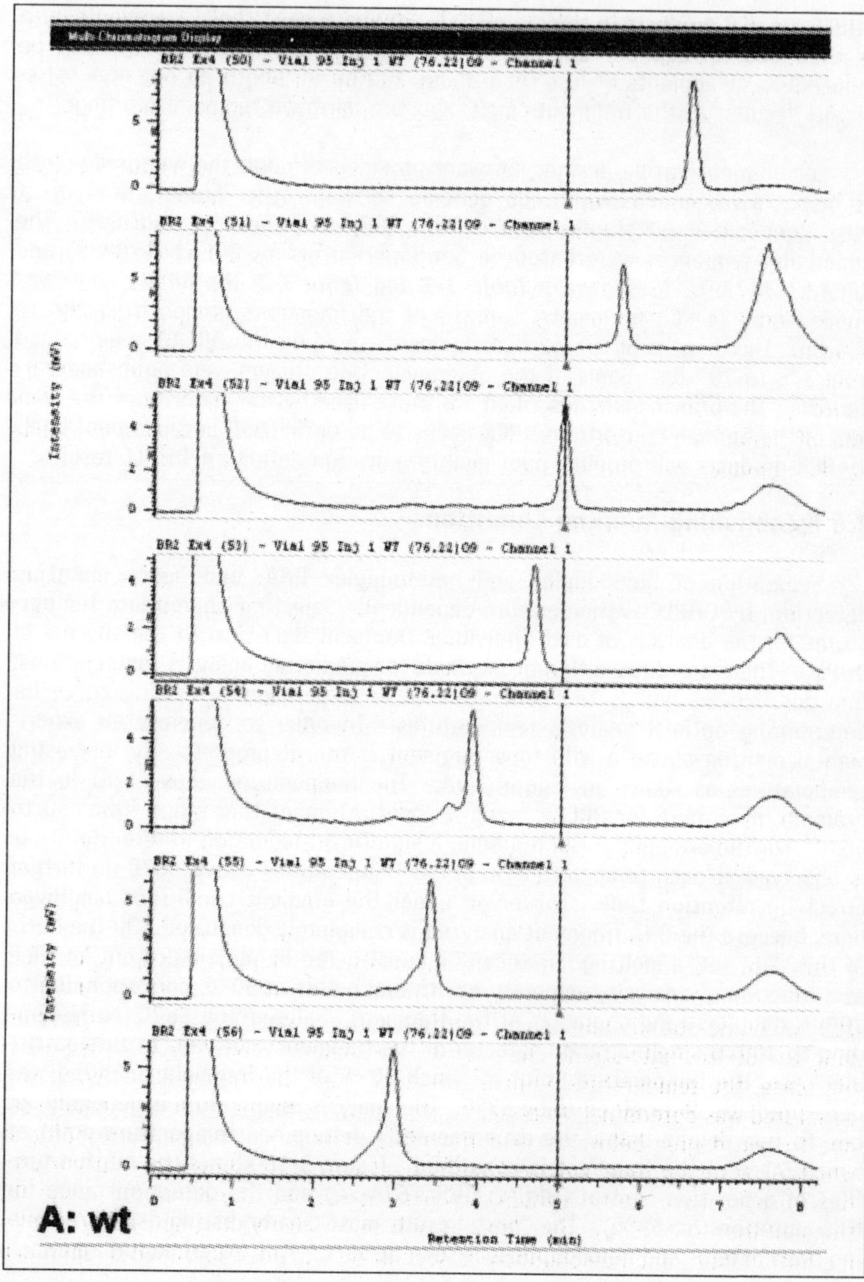
BR2 Ex4 (50) - Vial 95 Inj 1 WT (76.22)09 - Channel 1
BR2 Ex4 (51) - Vial 95 Inj 1 WT (76.22)09 - Channel 1
BR2 Ex4 (52) - Vial 95 Inj 1 WT (76.22)09 - Channel 1
BR2 Ex4 (53) - Vial 95 Inj 1 WT (76.22)09 - Channel 1
BR2 Ex4 (54) - Vial 95 Inj 1 WT (76.22)09 - Channel 1
BR2 Ex4 (55) - Vial 95 Inj 1 WT (76.22)09 - Channel 1
BR2 Ex4 (56) - Vial 95 Inj 1 WT (76.22)09 - Channel 1
Intensity (mV)
Retention Time (min)
A: wt

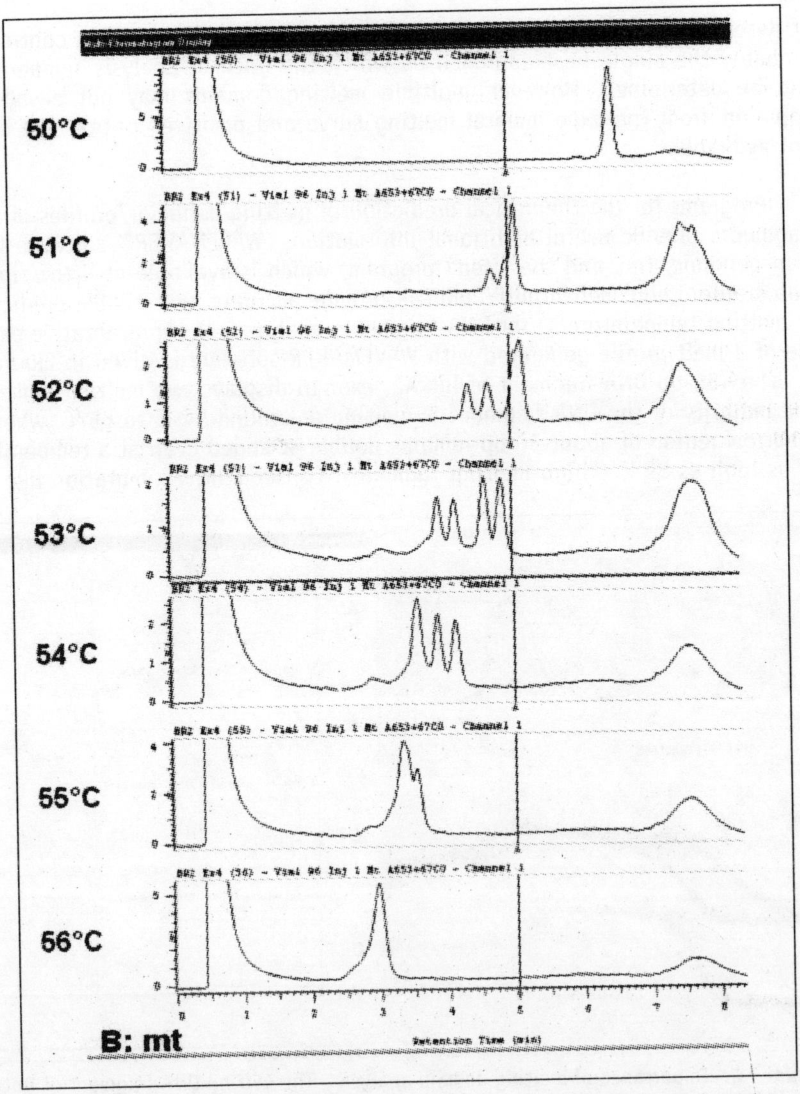

Figure 3-1: Generation of experimental melting curves. *(Opposite page)* (**A**) In order to generate an experimental melting curve, a wild type fragment (BRCA2, exon 4) was run at progressively increasing temperatures from 50°C to 56°C. Within this range a significant reduction of retention times can be observed at each temperature step. Further increase above 56°C has no effect because the DNA fragment is completely denatured. (**B**) Elution profiles of a corresponding positive control (BRCA2, IVS4+67A->C). The detection range for this mutation is between 51°C and 55°C. The best resolution/result, displaying both hetero- and homoduplices, is seen at 53°C.

of determining melting curves of wild type and respective positive controls is probably the single best approach by which the optimal analysis temperature can be determined. However, multiple melting domains may not always be apparent from the experimental melting curve and positive controls are often not available.

Programs for the theoretical prediction of melting behavior/profiles of DNA fragments provide useful additional information. WAVEMAKER™ software from Transgenomic, Inc. and the "Melt" program, which is available at http://insertion.stanford.edu/melt.htmlHYPERLINK provide accurate and reliable prediction of melting temperature(s) for DNA fragments analyzed. A representative example of a melt profile generated with WAVEMAKER software is given in Figure 3-2. The 450-bp DNA fragment of BRCA2, exon16 displays two melting domains. The majority of the DNA fragment is melting at around 54°C to 55°C, whereas a narrow region of about 30 bp remains double-stranded even at a temperature of as high as 59°C ("high-melting" domain). Consequently, a mutation like the

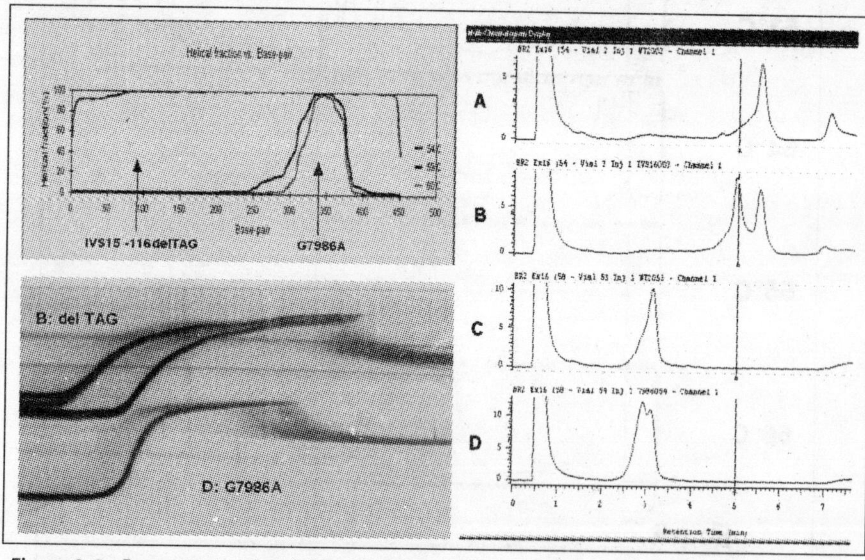

Figure 3-2: Importance of melting domain analysis. The 450-bp DNA fragment of BRCA2, exon16 displays two melting domains. Most of the DNA fragment starts melting at 54°C to 55°C. However, a "high-melting" domain of about 30 base pairs remains as double strand DNA even at a temperature of 59°C (upper left panel). Consequently, mutations like the IVS15-116delTAG located in the low-melting part of this DNA-fragment are detected at 54°C (right panel, B) whereas the mutation G7986A in the high-melting domain is seen only at 58°C (right panel, D). In the TGGE analysis the G7986A alteration could not be detected.

IVS15-116delTAG, which is located in the low-melting part of this DNA fragment, is detected at 54°C. In contrast, mutation G7986A, which is located in the high-melting domain, is seen only at 58°C. This example emphasizes the need for using theoretical melting curves predicted with WAVEMAKER software. Furthermore, it points out the superior sensitivity of the DHPLC method compared to the TGGE, where the G7986A alteration could not be detected (Figure 3-2). In most cases the aforementioned melt programs provide reliable analysis conditions allowing one to avoid the laborious generation of experimental melting curves. If possible, heteroduplex resolution should be additionally checked with positive controls that harbor known sequence variations.

The group in Düsseldorf collected a large panel of samples of different BRCA1/BRCA2 mutations, which are listed in the BIC database, for the German BRCA Consortium by asking the depositors into the database for DNA samples or PCR products. Due to the good feedback, the panel of positive controls includes at least one mutation for each BRCA1/BRCA2 fragment and in most cases for each melting domain. All positive controls were PCR amplified and PCR products were distributed to each center of the Consortium for establishing and validating DHPLC running conditions.

DHPLC analysis was carried out using a WAVE® Nucleic Acid Fragment Analysis System equipped with a DNASep® Cartridge (Transgenomic Inc., San Jose, CA). Crude PCR samples were injected after heteroduplex formation and eluted at a flow rate of 0.9 ml/min using a linear gradient of acetonitrile (ACN) in 0.1 M triethylammonium acetate (TEAA), pH 7.0. The software-predicted ACN gradient is adjusted in such a way that the elution time of the peaks falls within a 3- to 4-minute window. Generally, the analysis of an individual sample takes 8 - 10 min. This includes regeneration and equilibration of the column after each analysis. A representative elution gradient is shown in Figure 3-3. DHPLC running conditions for all amplicons of the BRCA1 and BRCA2 gene are summarized in ***Table 3-3*** and ***Table 3-4***. The analysis portion of each gradient is preceded by a loading drop of 0.5 min, e.g.: the PCR product to be analyzed is loaded onto the DNASep Cartridge at a percent Buffer B that is 5 % lower than the starting percentage of Buffer B of the analysis portion of the gradient. In order to maximize the sensitivity of the technology and detect all possible sequence variations, especially in fragments harboring different melting domains, it is recommended to perform mutation analysis at several different temperatures. In many cases more than one analysis temperature per amplicon is listed in tables 3 and 4. If the temperatures given differ by only one or two degrees, the second analysis temperature is generally required in order to distinguish between different sequence variants (see below) by DHPLC[8]. Therefore, a second run is necessary and is performed only in those cases that show a heteroduplex profile. For example, the run at 49°C with the BRCA1/exon

A

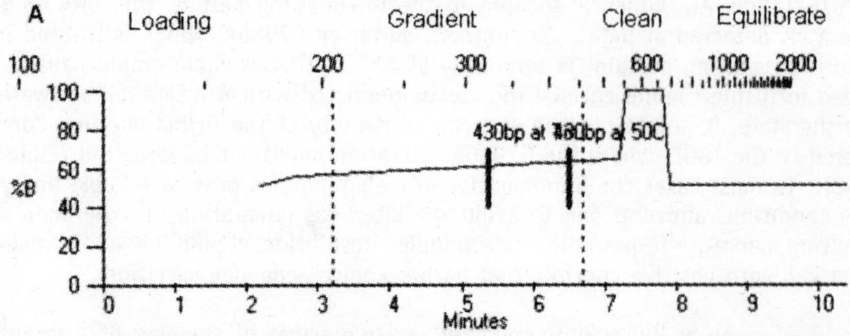

B

Step	Time	%A	%B	ml/min	At Detector
Loading	0.0	49	51	0.9	2.2
Start Gradient	0.5	44	56		2.7
Stop Gradient	5.0	35	65		7.2
Start Clean	5.1	0	100		7.3
Stop Clean	5.6	0	100		7.8
Start Equilibrate	5.7	49	51		7.9
Stop Equilibrate	8.2	49	51		10.4

Figure 3-3: Typical running conditions for DHPLC analysis on the WAVE Nucleic Acid Fragment Analysis System (Transgenomic Inc., San Jose, CA). Elution is performed at a flow rate of 0.9 ml/min using a linear gradient of buffer B (2%/min) [Buffer B: 25 % acetonitrile (ACN) in 0.1 M triethylammonium acetate (TEAA), pH 7.0]. The range of the ACN gradient is adjusted such that the elution of the peaks of interest occurs after 3 to 4 min into the run. Generally, the analysis of an individual sample takes 8-10 min. This includes regeneration and equilibration of the column after each analysis.

2 fragment is to distinguish between the profiles produced by the intronic variation (IVS1-115t>c) alone and in combination with the intronic sequence variant 185delAG. Analysis of the BRCA1/exon 11I fragment at two different temperatures permits differentiation between the sequence variants 3232A>G and 3238G>A based on their peak profiles. Due to observed variations in column oven performance, which will affect the reproducibility of DHPLC peak pattern, the conditions provided may need adjustment if transferred from one instrument to another. Therefore, in routine genetic diagnostics it is recommended to check established running conditions with a relevant panel of positive controls before each run.

3.7 DHPLC is More Than a Pre-screening Technique

DHPLC is typically used as a pre-screening technique, followed by DNA-sequencing for confirmation and identification of the sequence alteration leading to a DHPLC-positive result. In the case that amplified gene regions contain known, common polymorphisms the sequencing costs can markedly increase. If these harmless variations cannot be identified unambiguously, the advantage of a pre-screening method turns out to be of minor impact, because a large number of samples have to be sequenced in the end. This is the case for the BRCA1/BRCA2 genes where some fragments are highly polymorphic. In order to overcome this problem, two approaches have been followed.

(**A**) Elution profiles produced by individual sequence alterations, e.g. common polymorphisms, have been identified and characterized.

(**B**) DHPLC-positive samples are re-analysis after mixing equal quantities of the DHPLC-positive sample with a sample containing the sequence alteration in question.

Some of the known BRCA1/BRCA2 variants produce specific chromatogram profiles that differ from known polymorphisms also occurring in the same fragments[8]. In some cases the specific differences in elution profiles become most apparent if DHPLC analysis is conducted at different temperatures. Therefore, analysis of relevant DNA-fragments at different temperatures will increase reproducibility and reliability of characteristic mutant patterns (see above and Figure 3-4). However, the reliability of a qualitative DHPLC analysis has to be tested for each gene individually because some sequence regions may produce very similar or even identical DHPLC peak patterns for different variants. The importance of multiple temperature analysis as well as frequent column cleaning is emphasized. Successful separation of heteroduplices on the DNASep® cartridge is influenced by several factors. We observed that peak resolution might decrease with increasing age of the column. Consequently, an objective assessment of column performance becomes necessary. This is of particular importance for exons where few control samples are available that can be used to assess optimal column performance and differentiation between peak profiles. Several factors can influence column resolution, e.g. column batch, number of injections, additives in PCR reactions, or inadequate treatment of the column. Daily cleaning of the column with 75% acetonitrile at 60°C prolongs column performance tremendously. We have obtained up to 7000 injections on a single column using the resolution of the mutation standards from Transgenomic as a determinant for column performance and resolution. Nevertheless, it seems very difficult to judge if column performance is sufficient to predict specific sequence variations reliably. We propose to perform column performance test-

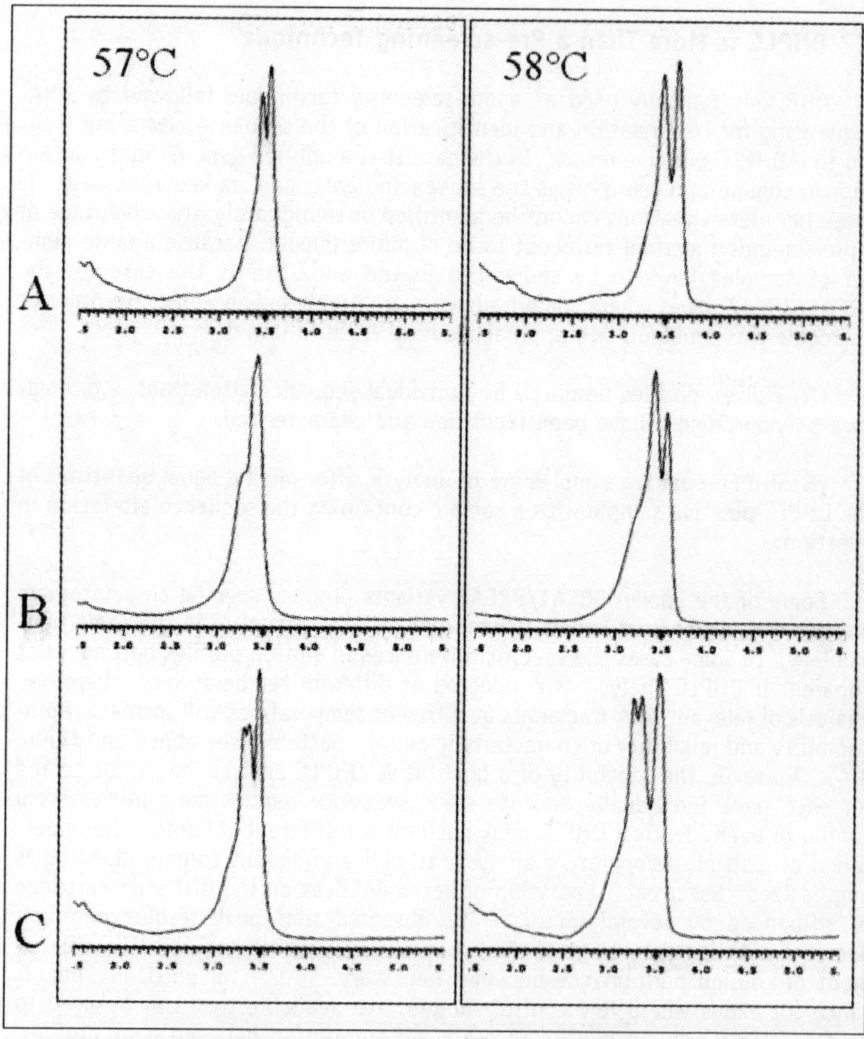

Figure 3-4: Discrimination between deleterious BRCA1 alterations and a known polymorphism by distinct elution profiles at two temperatures. Both BRCA1 variants shown in panel B and C produce specific chromatogram profiles different from the known A3667G polymorphism also occurring within the same DNA fragment (BRCA1, exon 11). The difference in elution profiles of the A3667G and the 3741insA mutation is more apparent at 57°C, the elution profile of the G3780T mutation clearly differs from both other mutations at both temperatures. BRCA1 exon 11, A: (3667 A>G(Lys 1183 Arg), B: (3741 ins A(1218 stop), C: (3780 G>T(Glu 1221 stop)

ing at least once per week, which may include running several samples with similar elution profiles. Hereby, the resolution of DHPLC peak patterns can be observed in consecutive runs. When test samples for column performance indicate altered heteroduplex resolution and cleaning does not improve the performance, the column should be changed.

A second alternative for confirming DHPLC-positive results of DNA fragments containing frequent polymorphisms is illustrated in Figure 3-5. Equal quantities of PCR product of unknown samples and a control DNA with a single known heterozygous sequence alteration are mixed, heated, and re-analyzed after renaturation using the same DHPLC conditions that were used to detect the sequence variation initially. The absence of change in the elution profile of the mixed sample clearly indicates that the mutation in the sample is the same as in the control. Mutations in the sample other than those contained in the control lead to distinct patterns comparable with elution profiles of samples containing two heterozygous mutations. This procedure requires the preparation of a panel of control samples and an additional DHPLC analysis for each positive DNA-fragment. Using the above approach, mutations or polymorphisms can be reliably identified and DNA-sequencing can be omitted.

3.8 Sensitivity of DHPLC and Quality Management of BRCA1/2 Mutation Analysis in the German Consortium for Hereditary Breast and Ovarian Cancer

The efficacy of DHPLC for mutational screening of the BRCA1 and BRCA2 genes is generally accepted. There are some papers in recent years documenting the sensitivity and specificity of DHPLC in mutation detection. In a recently published study comparing SSCP, CSGE, TDGS and DHPLC in BRCA1 mutation analysis, only the laboratory using DHPLC correctly identified all mutations[18]. Direct sequence analysis of 626 BRCA1 fragments previously subjected to DHPLC analysis confirmed the high sensitivity and specificity of DHPLC, both of which were 100%[7]. In a more recent report[8], DHPLC detected all of 432 heterozygotes in BRCA1 and 136 in BRCA2, with only one false positive. By comparing the chromatographic peak pattern of samples with that of standards containing known alternations, the identities of mutations were determined correctly in all but two cases without the need of sequencing[8]. In a separate study, out of 186 BRCA1 and BRCA2 mutations, only two sequence variations in exon 5 of BRCA1 yielded profiles difficult to distinguish from the chromatogram of a homozygous control[23]. The reason may lie in this particular fragment's ability to form a stable secondary cruciform structure, warranting its direct sequencing. But even if this and nine other BRCA1 amplicons that contain 16 frequent sequence variants were to be sequenced routinely to avoid missing mutations that yield DHPLC profiles similar to those of known polymorphisms, the cost

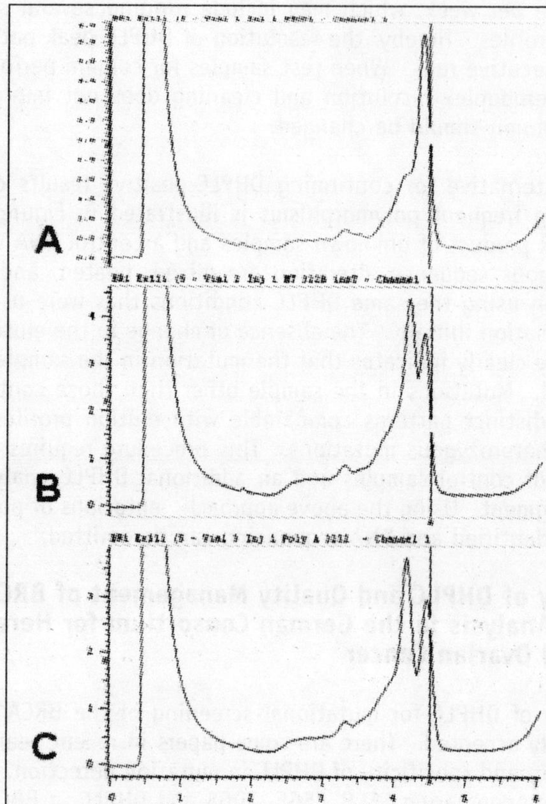

Figure 3-5: Re-analysis of an unknown DHPLC positive sample with a similar elution profile as a known polymorphism. Elution profiles of the BRCA1 exon11 fragment (**A**) wild type, (**B**) unknown alteration, (**C**) known polymorphism (A3232G), (**D**) unknown alteration + A3232G. Equal quantities of both PCR products were mixed, heated, and renatured. Analysis under the same DHPLC conditions showed a mutation profile clearly distinct from the profile of the A3232G polymorphism indicating a different alteration which was confirmed as a heterozygous 3238insT mutation by DNA sequencing.

savings obtained with DHPLC are significant due to the great number of amplicons (N = 34) required to cover the entire BRCA1 gene.

In conclusion, the sensitivity and specificity of DHPLC appear to be consistently higher than 96%. Not all failures can be readily explained. Challenging are mutations located in extremely high-melting domains surrounded by sequence that melts at significantly lower temperatures.

As mentioned above the German Consortium for Hereditary Breast and Ovarian Cancer started routine BRCA1/BRCA2 mutation analysis in 1997 and in 1999. Ten centers of the consortium established simultaneously the DHPLC methodology for germline mutation screening in the BRCA1/BRCA2 genes on a WAVE Nucleic Acid Fragment Analysis System (Transgenomic, Inc.). In order to reach and maintain a high level of quality of genetic testing for BRCA genes the consortium members decided to implement a program for external quality assessment (EQA). For this purpose, an external expert in the field provided a BRCA1/BRCA2-specific EQA scheme. DNA samples were sent out with mock clinical referrals and a request for diagnosis. Written reports, including genotyping results and interpretations, had to be returned. The aim of this evaluation was to measure the actual, average performance level and to estimate the quality of testing performed by individual laboratories.

A set of 9 genomic DNA samples with BRCA1/BRCA2 mutations was sent out and 9 exons for each sample had to be analyzed. Results, including sequence alteration identified and their classification, had to be reported. In total 410 BRCA1/BRCA2 amplicons were analyzed. In 5 cases (1.22%) false negative DHPLC results were obtained. The error rate for the identification and classification of DHPLC positive results was 1,73 %. Each false negative DHPLC result occurred only once per exon and the mutation was detected in the other 8 laboratories correctly. Moreover 4/5 errors occurred in 2 centers indicating that DHPLC sensitivity in general is above 98 %. We conclude that the accuracy and sensitivity of DHPLC for mutation detection is in part dependent on the skill level and experience in interpreting elution profiles of the person(s) performing the analysis. While an accuracy of over 98 % may appear good at first glance, it must be kept in mind that much of the BRCA1/BRCA2 testing is done as predictive testing on individuals who may have an increased family risk but are free of the cancer disease. Clearly, predictive testing needs the highest possible reliability.

3.9 Results From the German Consortium for Hereditary Breast and Ovarian Cancer

The main focus of this German-wide, multi-center study was to establish a BRCA1/2 mutation profile and to determine family types with high frequencies of mutations in these genes. In a comprehensive study, the entire coding sequences of the breast cancer genes BRCA1 and BRCA2 were analyzed in 989 unrelated patients from German breast/ovarian cancer families.[21] A total of 77 BRCA1 and 63 BRCA2 distinct deleterious mutations were found in 302 patients. More than 1/3 of these mutations are novel and might be specific to the German population. Eighteen common mutations were found in 68% of cases in BRCA1 and 13 recurrent mutations in 44% of cases in BRCA2, facilitating prescreening approaches. Haplotype analyses indicate that 14 out of 20 recurrent mutations are likely originating from a common founder.

An additional 50 different rare sequence variants with unknown relevance for tumor genesis were found in 72 families. Correlation of BRCA1/BRCA2 detection rates with family history identified families with both breast and ovarian cancer to be at highest risk for BRCA1/BRCA2 mutations (43% and 10%, respectively), followed by families with at least 2 pre-menopausal cases of breast cancer (24% BRCA1 and 13% BRCA2 mutations). These data provide strong evidence for further predisposing genes in the German population. In breast cancer families with 2 or 3 affected females but only a single or no pre-menopausal case, mutations were detected with low frequencies (about 10% or less for both genes). The decision for or against molecular diagnosis is now aided by considering the expected mutation detection rates that greatly depend on family history and structure.

Table 3-1: Primer sequences and PCR conditions of BRCA1

Exon	5′-Primer (forward)	3′-Primer (reverse)	Size (bp)	Annealing Temp.(°C)	$MgCl_2$ (mM)
2	aggacgttgtcattagttctttg	aaggtcaattctgttcatttgc	319	58	4
3	cagttcctgacacagcagaca	ttcctgggttatgaaggacaaa	286	58	2
5	gcttgtaattcacctgccat	ttcctactgtggttgcttcc	269	58	2
6	aggttttctactgttgctgcat	aaaaggtcttatcaccacgtca	307	55	2
7	gggtttctcttggtttctttg	gaagaagaagaaaacaaatggt	251	TD* 55-48	2
8	agctgactgatgatggtcaa	aaattcacttcccaaagctg	338	55	2
9	tgccacagtagatgctcagt	cacatacatccctgaacctaaa	300	57	2
10	ttggtcagctttctgtaatcg	ccataccacgacatttgaca	346	55	2
11A	tagccagttggttgatttcc	cccatctgttatgttggctc	394	58	2
11B	ccatgtggcacaaatactca	tgattcagactccccatcat	399	58	2
11C	gaaactgccatgctcagaga	atttatttgtgaggggacgc	437	58	2
11D	tccccaacttaagccatgta	agaagacttcctcctcagcc	437	58	2
11E	ttcaaaacgaaagctgaacc	ttggaaggctaggattgaca	445	58	2
11F	ggtaaagaacctgcaactgg	tcaaatgctgcacactgact	416	58	2
11G	gaaagggttttgcaaactga	ttcctcttctgcatttcctg	381	58	2
11H	tgaacttgatgctcagtatttgc	agtccagtttcgttgcctct	345	56	2
11I	taagccagttgataatgcca	ttttggccctctgtttctac	430	60	2
11J	actaatgaagtgggctccag	caggtcatcaggtgtctcag	309	59	2
11K	ttcctggaagtaattgtaagca	taaccctgagccaaatgtgtat	313	60	2
11L	gacattaaggaaagttctgctg	tttgccaatattacctggttac	329	56	2
11M	accgttgctaccgagtgtct	gtgctccccaaaagcataaa	438	55	2
12	gtcccaaagcaaggaattta	tcaaagagatgatgtcagcaa	327	54	2
13	ggtgatttcaattcctgtgc	aaatgttggagctaggtccttac	373	58	2
14	tgtgtatcatagattgatgcttttg	gcaataaaagtgtataaatgcctgt	361	60	4
15	ttgccagtcatttctgatct	aaaccttgattaacacttgagc	482	54	2
16	cactgtattcatgtacccattt	cacagaactgtgattgttttct	564	55	2
17	tgagctgtgtgctagaggtaa	ctcgcctcatgtggttttat	202	57	2
18	agtggtgttttcagcctctg	ctcagactcagcatcagcaa	342	58	2
19	ttaaagggctgtggctttag	aaggaaagtggtgcattgat	279	54	2
20	tgtctgctccacttccattga	tacaaaatgaagcggcccat	246	59	2
21	caggtggtgaacagaagaaa	acatttcagcaatctgagga	298	54	2
22	catccggagagtgtagggta	catccatagggactgacagg	240	60	2
23	ccctgtctcaaaaacaaaca	caagcaccaggtaatgagtg	234	57	3
24	tggagtcgattgattagagc	agccaggacagtagaaggac	311	60	2

*TD = touchdown PCR

Table 3-2: Primer sequences and PCR conditions of BRCA2

Exon	5′-Primer (forward)	3′-Primer (reverse)	Size (bp)	Annealing Temp.(°C)	$MgCl_2$ (mM)
2	tgttcccatcctcacagtaag	gtactgggtttttagcaagca	336	58	3
3	aactgttctgggtcacaaat	agaggccagagagactgatt	439	57	2
4	aacacttccaaagaatgcaa	taccaggctcttagccaaa	292	57	2
5 and 6	tggcattttaaacatcacttg	ctcagggcaaaggtataacg	450	58	2.5
7	agcattctgcctcatacagg	tcaacctcatctgctctttctt	284	58	2.5
8	tcactgtgttgattgacctttc	ggcattccaaaattgttagc	278	58	2.5
9	ggacctaggttgattgcaga	caacaacaacaaaaaaacctg	280	TD*63-56	3
10.1	tataaaatattaatgtgcttctgtt	aaagggcttctgatttgctac	374	TD63-56	2.5
10.2	gcaaacgctgatgaatgtg	gaaatgaagaagccactgga	490	57	2
10.3	agcagcatcttgaatctcat	ggtacctgaatcagcatttg	503	57	2
11A	aagcagtcttcctgcctcag	gcagccaagacctcttctttt	446	60	2
11B	ccccagaagctgattctctgt	tttcaggtggcaacagctc	405	TD63-58	3.5
11C	tcccatggaaaagaatcaag	cctctgcaagaacataaacca	422	54	2
11D	cgaacccattttcaagaact	ggcttgctcagtttcttttg	432	58	2.5
11E	ggaaatcaagctctctgaaca	attgcttgctgctgtctacc	500	57	2
11F	aagtgcctgaaaaccagatg	caacaaaagtgccagtagtca	492	TD63-58	3.5
11G	gtcatgattctgttgtttca	gactctttggcgacactaat	532	56	2
11H	tgctactaaaacggagcaaa	ggtctttacaggcctctctg	497	TD65-61	3.5
11I	ggaatctttggacaaagtga	ggttgaccatcaaatattcc	498	55	2
11J	tcagtccccttattcagtca	tgcagggtgaagagctagt	492	58	2
11K	cttggattctggtattgagc	cactctgaatgtcagcaaaa	479	55	2
11L	aacgaaaattatggcaggttgttac	gctttccacttgctgtactaaatcc	507	TD68-61	2
11M	attcagaccagctcacaaga	aggtgaagcctgttcttttc	500	54	2
11N	tgttgaaggtggttcttcag	actttaaaaatagtgattggcaac	480	55	3
12	tgtggtatctggtagcatctg	cacagtggctcatgtctgta	494	60	2
13	gtgagttatttggtgcatagtc	cgagacttttctcatactgtattag	317	57	2.5
14.1	ggctagccttgaaaaatgtg	aaagactttggttggtctgc	382	58	2.5
14.2	gcaatttagcagtttcaggac	gggctttaaaattaccacca	428	58	2.5
15	ggccaggggttgtgcttttt	atttcattcatccattcctgc	368	TD63-56	2.5
16	gcagactgtggaatgtatgg	agaaagagggatgagggaat	452	57	2
17	cagagaatagttgtagttgttgaa	agaaaccttaacccatactgc	306	TD63-56	2.5
18	tcagtgacttgtttaaacagtg	catctaagaaattgagcatcc	537	54	2
19	aaggcagttctagaagaatgaaa	caagagaccgaaactccatc	342	58	2
20	ggtgatccactaatctcagc	tgtcccttgttgctattctt	452	57	2
21	ctttgggtgttttatgcttg	catattccttcctgtgatgg	300	58	3

Table 3-2: Primer sequences and PCR conditions of BRCA2*(continued)*

Exon	5'-Primer (forward)	3'-Primer (reverse)	Size (bp)	Annealing Temp.(°C)	$MgCl_2$ (mM)
22	tttgtcctgtttaaagccatc	agtggattttgcttctctga	409	57	2
23	ccactactaatgcccacaaa	caagcacttatcaaaactgaaa	324	57	2
24	accggtacaaacctttcatt	aatttgccaactggtagctc	334	57	2
25	agcactgtaagcaacaggtc	taccaaaatgtgtggtgatg	463	57	2
26	ggtcccaaacttttcatttc	agaatatacgatggcctcca	310	57	2
27.1	ctgtgtgtaatatttgcgtgct	tcaatgcaagttcttcgtca	500	57	2
27.2	ccaaatacgaaacacccata	cgctgaggtaaatttgaaac	476	57	2

*TD = touchdown PCR

Table 3-3: DHPLC running conditions for BRCA1

Exon	DHPLC gradient (%B within 3min)	DHPLC Temperature (°C)	Exon	DHPLC gradient (%B within 3min)	DHPLC Temperature (°C)
2	I. 59-65%B* II. 57-63%B II. 60-66%B	52 56 49	3	I. 58-64%B II. 57-63%B	56 57
5#	I. 58-64%B II. 58-64%B	54 55	6	58-64%B	57
7	59-65%B	56	8	I. 57-63%B II. 57-63%B	59 58
9	I. 59-65%B II. 59-65%B	54 53	10	I. 59-65%B II. 60-66%B	56 54
11A	59-65%B	58	11B	I. 60-66%B II. 59-65%B	56 57
11C	I. 62-68%B II. 60-66%B	51 56	11D	60-66%B	56
11E	59-65%B	56	11F	60-66%B	56
11G	I. 60-66%B II. 58-64%B	56 58	11H	59-65%B	57
11I	I. 60-66%B II. 58-64%B	55 57	11J	60-66%B	55
11K	58-64%B	57	11L	58-64%B	58

It is recommended that exon 5 be sequenced.

Table 3-3: DHPLC running conditions for BRCA1 *(continued)*

Exon	DHPLC gradient (%B within 3min)	DHPLC Temperature (°C)	Exon	DHPLC gradient (%B within 3min)	DHPLC Temperature (°C)
11M	I. 60-66%B II: 58-64%B	57 59	12	I. 54-60%B II. 52-58%B	60 61
13	I. 58-64%B II. 60-66%B	60 55	14	I. 58-64%B II. 57-63%B	55 57
15	I. 58-64%B II. 60-66%B	58 55	16	I. 56-62.2%B II. 58-64%B	60 56
17	58-64%B	57	18	59-65%B	56
19	I. 60-66%B II. 58-64%B	55 58	20	I. 58-64%B II. 59-65%B	58 55
21	I. 57-63%B II. 59-65%B	60 56	22	I. 58-64%B II. 60-66%B	59 56
23	I. 57-63%B II. 56-62%B	61 62	24	I. 56-62%B II. 59-65%B	62 56

*Eluent B was 0.1 M Triethylammonium acetate (TEAA), 25% acetonitrile (ACN)

Table 3-4: DHPLC running conditions for BRCA2

Exon	DHPLC gradient (%B within 4min)	DHPLC Temperature (°C)	Exon	DHPLC gradient (%B within 4min)	DHPLC Temperature (°C)
2	I. 57-63%B* II. 58-64%B	58 56	3	I. 56-65%B II. 57-66%B	57 56
4	I. 59-68%B II. 57-66%B	53 55	5+6	I. 58-67%B II. 56-65%B	52 54
7	56-65%B	57	8	I. 56-65%B II. 57-66%B	54 53
9	I. 56-65%B II. 57-66%B	56 54	10.1	I. 56-65%B	54
10.2#	56-65%B	56	11A	I. 56-65%B II. 57-66%B	55 54
11B	I. 55-64%B II. 56-65%B	57 55	11C	I. 58-67%B II. 57-66%B	53 54

Table 3-4: DHPLC running conditions for BRCA2 *(continued)*

Exon	DHPLC gradient (%B within 4min)	DHPLC Temperature (°C)	Exon	DHPLC gradient (%B within 4min)	DHPLC Temperature (°C)
11D	I. 58-67%B II. 60-69%B	55 53	11E	54-63%B	56
11F	I. 58-67%B II. 56-65%B	53 58	11G	I. 58-67%B II. 58-67%B	54 53
11H	I. 54-63%B II. 52-61%B	56 58	11I	I. 54-63%B II. 51-60%B	55 58
11J	I. 58-67%B II. 55-64%B	53 56	11K	55-64%B	55
11L	I. 56-65%B II: 52-61%B	54 58	11M	58-67%B	54
11N	I. 56-65%B II. 54-63%B	55 57	12	I. 58-67%B II. 56-65%B	52 54
13	I. 58-67%B II. 59-68%B	53 51	14.1	54-63%B	54
14.2	I. 55-64%B II. 56-65%B	55 53	15	56-65%B	55
16	I. 56-65%B II. 54-63%B	54 57	17	I. 54-63%B II. 54-63%B	56 57
18	I. 53-62%B II. 57-66%B	58 52	19	I. 54-63%B II. 57-66%B	57 52
20	I. 54-63%B II. 57-66%B	57 52	21	I. 53-62%B II. 56-65%B III.52-61%B	58 54 60
22	I. 54-63%B II. 53-62%B III. 68-67%B	56 58 53	23	56-65%B	55
24	I. 56-65%B II. 58-67%B III. 54-63%	55 53 58	25	I. 54-63%B II. 55-64%B	58 56
26	54-63%B	57	27.1	I. 56-65%B II. 54-63%B	54 59
27.2	I. 56-65%B II. 56-65%B	55 54			

* Eluent B was 0.1 M Triethylammonium acetate (TEAA), 25% acetonitrile (ACN)

Fragment 10.3 sequencing is recommended

References

1. ***Miki Y, Swenson J, Shattuck-Eidens D, Futreal PA, Harshman K, Tavtigian S, Liu K, Cochran C, Bennett LM, Ding W et al:*** (1994). A strong candidate for the breast and ovarian cancer susceptibility gene BRCA1. *Science* 266: 66-71.

2. ***Wooster R, Neuhausen SI, Mangion J, Quirk J, Ford D, Collins N, Nguyen K, Tran T, Averill D et al:*** (1994). Localization of a breast cancer susceptibility gene, BRCA2, to chromosome 13q12-13. *Science* 265: 2088-2090.

3. ***Ford D, Easton D F, Stratton M, Narod S, Goldgar D, Devilee P, Bishop D T, Weber B, Lenoir G, Chang-Claude J, Sobol H, Teare M D, and 27 others:*** (1998). Genetic heterogeneity and penetrance analysis of the BRCA1 and BRCA2 genes in breast cancer families. *Am. J. Hum. Genet.* 62: 676-689

4. ***Frank TS, Manley SA, Olopade OI, Cummings S, Garber JE, Bernhardt B, Antman K, Russo D, Wood ME, Mullineau L, Isaacs C, Peshkin B, Buys S, Venne V, Rowley PT, Loader S, Offit K, Robson M, Hampel H, Brener D, Winer EP, Clark S, Weber B, Strong LC, Thomas A, et al:*** (1998) Sequence analysis of BRCA1 and BRCA2: correlation of mutations with family history and ovarian cancer risk. *J Clin Oncol* 16(7):2417-25

5. ***Frank TS, Deffenbaugh AM, Reid JE, Hulick M, Ward BE, Lingenfelter B, Gumpper KL, Scholl T, Tavtigian SV, Pruss DR, Critchfeld GC:*** (2002). Clinical characteristics of individuals with germline mutations in BRCA1 and BRCA2: analysis of 10,000 individuals. *J Clin Oncol 20*(6):1480-90

6. ***Nicoletto MO, Donach M, De Nicolo A, Artioli G, Banna G, Monfardini S:*** (2001). BRCA-1 and BRCA-2 mutations as prognostic factors in clinical practice and genetic counselling. *Cancer Treat Rev* 27: 295-304.

7. ***Arnold N, Gross E, Schwarz-Boeger U, Pfisterer J, Jonat W, Kiechle M:*** (1999). A highly sensitive, fast, and economical technique for mutation analysis in hereditary breast and ovarian cancers. *Hum Mutat* 14: 333-339.

8. ***Gross E, Arnold N, Pfeifer K, Bandick K, Kiechle M:*** (2000). Identification of specific BRCA1 and BRCA2 variants by DHPLC. *Hum Mutat* 16: 345-353.

9. ***Hogervorst FB, Cornelis RS, Bout M, van Vliet M, Oosterwijk JC, Olmer R, et al:*** (1995) Rapid detection of BRCA1 mutations by the protein truncation test. *Nat Genet* 10:208-212. Breast Cancer Information Core, BIC website: (http://research.nhgri.nih.gov/bic/).

10. ***Markoff A, Savov A, Vladimirov V, Bogdanova N, Kremensky I, Ganev V:*** (1997) Optimization of single-strand conformation polymorphism analysis in the presence of polyethylene glycol. *Clin Chem* 43:30-33.

11. ***Lancaster JM, Berchuck A, Futreal PA, Wiseman RW:*** (1997) Dideoxy fingerprinting assay for BRCA1 mutation analysis. *Mol Carcinog* 19:176-179.

12. ***Fodde R, Losekoot M:*** (1994) Mutation detection by denaturing gradient gel electrophoresis (DGGE). *Hum Mutat* 3:83-94.

13. ***van Orsouw NJ, Dhanda RK, Elhaji Y, Narod SA, Li FP, Eng C, Vijg J:*** (1999). A highly accurate, low cost test for BRCA1 mutations. *J Med Genet* 36: 747-753.

14. ***Markoff A, Sormbroen H, Bogdanova N, Preisler-Adams S, Ganev V, Dworniczak B, Horst J:*** (1998) Comparison of conformation-sensitive gel electrophoresis and single-strand conformation polymorphism analysis for detection of mutations in the BRCA1 gene using optimized conformation analysis protocols. *Europ J Hum Genet* 6:145-150.

15. ***Taylor GR and Deeble J:*** (1999) Enzymatic methods for mutation scanning. *Genet Anal* 14:181-186.

16. ***Richter S, Seth A:*** (1998) One step direct detection of recurrent mutations in the breast cancer susceptibility gene BRCA1. *Int J Oncol* 12:1263-1267.

17. ***Tonisson N, Kurg A, Kaasik K, Lohmussaar E, Metspalu A:*** (2000). Unravelling genetic data by arrayed primer extension. *Clin Chem Lab Med* 38: 165-170.

18. ***Eng C, Brody LC, Wagner TM, Devilee P, Vijg J, Szabo C, Tavtigian SV, Nathanson KL, Ostrander E, Frank TS:*** (2001). Interpreting epidemiological research: blinded comparison of methods used to estimate the prevalence of inherited mutations in BRCA1. *J Med Genet* 38: 824-833.

19. ***Underhill PA, Jin L, Lin AA, Mehdi SQ, Jenkins T, Vollrath D, Davis RW, Cavalli-Sforza LL, Oefner PJ:*** (1997). Detection of numerous Y chromosome biallelic polymorphisms by denaturing high-performance liquid chromatography. *Genome Res* 7: 996-1005.

20. ***Xiao W, Oefner PJ:*** (2001). Denaturing high-performance liquid chromatography: A review. *Hum Mutat* 17: 439-474.

21. ***Meindl A on the behalf of the German Consortium for Hereditary Breast and Ovarian Cancer:*** (2002). Comprehensive analysis of 989 patients with breast or ovarian cancer provides BRCA1 and BRCA2 mutation profiles and frequencies for the German population. *Int J Cancer* 97: 472-480.

22. ***Ellis LA, Taylor CF, Taylor GR:*** (2000). A comparison of fluorescent SSCP and denaturing HPLC for high throughput mutation scanning. *Hum Mutat* 15: 556-564.

23. ***Wagner TMU, Stoppa-Lyonnet D, Fleischmann E, Muhr D, Pagès S, Sandberg T, Caux V, Moeslinger R, Langbauer G, Borg A, Oefner P:*** (1999). Denaturing high performance liquid chromatography (DHPLC) detects reliably BRCA1 and BRCA2 mutations. *Genomics* 62:369-376.

CHAPTER 4

MUTATION DETECTION IN THE TUBEROUS SCLEROSIS GENES TSC1 AND TSC2 BY DHPLC

Penelope S. Roberts, Sandra L. Dabora, David J. Kwiatkowski

Contact:
Dr. David J. Kwiatkowski
Hematology Division, Department of Medicine, Brigham and Women's Hospital,
Harvard Medical School, Boston, MA 02115
E-mail: dk@rics.bwh.harvard.edu

0-9664027-7-4/03/$0.00+$.50 *From:* **Genetic Variance Detection~Nuts & Bolts of DHPLC in Genomics** (pp.79-92)
Edited by: Karl H. Hecker, Ph.D.

4.1 Overview of Tuberous Sclerosis (TSC)

TSC is a relatively variable clinical syndrome, in which several pathologic manifestations are highly specific for the disease[1]. The cardinal feature is the cortical tuber, which is a distinctive form of brain hamartia. Hamartias and hamartomas usually involve several different organ systems in TSC patients, in addition to the brain. Involvement of the skin, heart, and kidney are particularly frequent, occurring at some point during the lifetime of most patients. A hamartia is a group of dysplastic, disorganized cells within an organ, but occurring without unusual growth. A hamartoma is the occurrence of cells in a similar aberrant pattern, but with some growth tendency or potential.

TSC occurs as an autosomal dominant disorder. Large families with the condition are rare due to reduced reproductive potential among those affected, related to the occurrence of seizures, mental retardation, and significant developmental delay. However, improvements in clinical care have led to increasing numbers of small families. About two-thirds of TSC cases are sporadic, due to the occurrence of new mutations. TSC occurs in up to 1 in 6,000 live births without ethnic clustering. Linkage of TSC to chromosome 9q34 was achieved in 1987, and the corresponding TSC1 gene was identified in 1997[2]. Linkage of TSC to chromosome 16p13 was discovered in 1992, and the corresponding TSC2 gene identified in 1993[3]. Among families large enough to permit linkage analysis, approximately half show linkage to 9q34 and half to 16p13[4]. There is no evidence for a third locus.

4.2 Gene Structure of TSC1 and TSC2

The TSC1 gene consists of 23 exons, of which 21 encode hamartin, with an 8.6-kb mRNA including a 4.5-kb 3′ untranslated region. There is variable splicing of the 2nd, untranslated exon. The gene occupies a genomic extent of 55 kb on 9q34. The TSC2 gene consists of 42 exons, of which 41 encode tuberin, with a 5.4-kb mRNA and relatively short 5′ and 3′ untranslated regions. There is alternative splicing of exon 25, the first 3 bp of exon 26, and exon 31. Tuberin derived from mRNA isoforms missing exons 25 and 31 is relatively common, so that the functional importance of regions encoded by those exons appears minimal[4].

4.3 Mutational Analysis in TSC1 and TSC2- Strategic Considerations

Several years ago when the mutational spectrum in the TSC1 and TSC2 genes was unknown, the best mutational analysis strategy was entirely uncertain. Well over 600 mutations in the TSC1 and TSC2 genes have now been identified (http://expmed.bwh.harvard.edu/ts/review/), and they comprise a diverse set of inactivating mutations (see below). It is clear that mutations can occur in nearly any of the exons of each of these genes, and that although there are 'warmspots' for mutation, there are no major hotspots for mutation that might have permitted a more focused mutation identification strategy. It is also now clear that large deletions (>1kb) and genomic alterations account for at least 10% of all mutations occurring in these genes, particularly TSC2, so that any mutation identification strategy must include a method for identification of these mutations.

Numerous methods have been developed to scan amplified DNA fragments for genetic variation/mutation. In our laboratory, we have experience with five methods: heteroduplex analysis by conformation sensitive gel electrophoresis (CSGE) on polyacrylamide gels, single-stranded conformation gel analysis (SSCP), denaturing gradient gel electrophoresis (DGGE), denaturing high pressure liquid chromatography (DHPLC), and resequencing. The first four of these methods identify amplicons that are likely to contain sequence variants, but sequencing must be performed for confirmation and identification of the precise variant.

The major focus of this chapter is to review our experience with DHPLC, but we briefly discuss our experience with these alternative methods. CSGE is an attractive method in which heteroduplexes are formed after PCR, and are identified by polyacrylamide gel electrophoresis performed under partially denaturing conditions to provide the maximum discrimination between heteroduplexes and homoduplexes. This method works extremely well for detection of insertion and deletion variants. Its sensitivity in detection of single nucleotide polymorphism (SNP) or mutation variants is debated. In our hands, it was much less than 100% sensitive (see below). It has been adapted for use of fluorescently labeled primers and analysis on ABI polyacrylamide gels and capillaries[5], and continues in use in some laboratories. SSCP analysis is one of the oldest methods available for variation analysis. Although it continues to have adherents, and there are publications describing a sensitivity of over 95% for SNP detection when performed under a series of different conditions[6], our experience using it under more conventional conditions is that it is much less than 90% sensitive (see below). DGGE is another time-honored technique, which has been adapted to a multiplex format[7]. This method is extremely sensitive when performed properly, achieving SNP detection rates well over 90%. However, implementation of the method in multiplex format is difficult and time-consuming. Routine operation of the method is also complicated by partial assay failure. In addition, the equipment needed for operation of this method is non-standard and moderately expensive.

Resequencing analysis is increasingly popular as a method for identification of sequence variation. In high throughput genome-style laboratories with good bioinformatics support, this method is relatively efficient. In our limited experience, the method is both expensive and time-consuming, in that the analysis of the sequence traces takes much time even for a skilled investigator or technician. Expense is considerably higher than any other method in which a mutation screening method is used followed by directed sequencing. However, as sequencing costs continue to fall, this approach becomes increasingly attractive. The infrastructure required in terms of sequencing capacity and bioinformatics is however significant. In addition, it is clear that resequencing is not 100% effective in detection of variation at the heterozygote level, and will miss many cases of mosaicism (see below).

4.4 Implementation of DHPLC Screening for Identification of Small Mutations in TSC1 and TSC2

DHPLC analysis of heteroduplexes to detect mutations and sequence variation has been used in our laboratory since 1999. Full details on the theory, mechanics, and details of the buffers and column matrix are given elsewhere in this book. The method is attractive because of the semi-automated nature of the analysis, so that 96 amplicons can be screened (14 to 18 hours per 96 well plate) easily during a 24-hour period in the standard analysis mode.

When we initially obtained our first machine, a WAVE DNA Fragment Analysis System (Transgenomic, Inc., Omaha, NE), there was relatively little guidance available for implementation. We designed primers so that exon amplicons would have a flat melting curve (http://insertion.stanford.edu/melt.html), adding GC clamps (4 - 10b) to primers to make the melting curve have a variation of no more than 4°C over the length of the amplicon[8]. Amplicons were chosen to amplify all exons of TSC1 and TSC2, including at least 20b of 5′ flanking and 10b of 3′ flanking sequence. For each amplicon, the run conditions on the DHPLC machine were determined empirically as follows. We used the Hansen and Oefner program (http://insertion.stanford.edu/melt.html) to determine a starting temperature for the DHPLC analysis. Temperatures within a range of 2 degrees up or down from this temperature were then investigated, and the optimal temperature was chosen based upon three criteria: first, the DNA elution peak should be shifted by at least one minute to the left from the elution peak when run at 55°C; second, the highest temperature possible was used at which a homozygous sample eluted as a sharp peak; and third, the temperature should give the best resolution of heterozygous and homozygous samples, when samples with known mutations were available.

Although there have been some pitfalls in the use of the DHPLC machines, in terms of proper care and consistent performance, this approach has become and remained the standard approach for detection of small mutations in TSC1 and TSC2 in our laboratory. The primers and methods have been modified, taking account of improved genomic sequence data for these genes. Periodically, we have also reexamined the method sets for each of the amplicons. For example, some time ago we performed an analysis of the run temperatures derived from the Hansen and Oefner program, those derived from a proprietary program offered by Transgenomic, and our working methods. We found that our oven temperatures were always within one or two degrees of the Hansen/Oefner program, but were generally 2-4°C hotter than those suggested by the Transgenomic software. We also tried out varying our temperatures for DHPLC runs of amplicons from multiple DNA samples for which mutations in TSC1 or TSC2 had not been found. This exercise proved fruitless, and therefore we have continued to use our empirically derived temperatures. In addition, a blinded comparison showed that our DHPLC run conditions were capable of detecting nearly all sequence variation detected by other mutation screening procedures (see below).

The elution profile of each amplicon from the DHPLC machine is unique. Many are single sharp peaks but they can also be broad or have a shoulder. The development of software to permit discrimination of variants among the elution profiles would be beneficial. Consequently, the human eye is at present the best detector of shifts that indicate the presence of variation. Printouts of elution profiles stacked one on top of the next facilitates visual discrimination of variant elution profiles.

4.5 Comparison of DHPLC with CSGE and SSCP

We performed a blinded comparison among three methods of mutation detection: CSGE, SSCP, and DHPLC. CSGE and DHPLC were applied in the analysis of the first 20 exons of TSC2 in 84 patients. All sequence variants detected by CSGE were identified by DHPLC, consisting of 7 distinct insertion or deletion variants and 13 distinct single base substitutions. DHPLC analysis identified an additional 14 distinct SNPs, so that CSGE had detected only 15 of 28 (54%) SNPs overall[8]. In a second comparative trial we analyzed 15 DNA samples in which 18 distinct sequence variants had been detected by SSCP. In the blinded DHPLC analysis, all 18 variants were detected, and an additional 7 distinct sequence variants were identified. Thus, overall 11 of 16 (69%) SNPs were identified by SSCP in these samples, while all were identified by DHPLC[8].

In a third comparative trial, sequence variants in the TSC1 gene that had been identified by a variety of techniques (CSGE, SSCP, and DGGE) were analyzed by DHPLC[9]. Twenty-seven of 28 (96%) sequence variants were identified by DHPLC. The one that was missed was present as a mosaic mutation, and had been detected by CSGE. Four or more additional distinct sequence variants were identified by DHPLC in these samples, and all of these had been missed by the earlier techniques.

4.6 Current Results of Mutational Analysis in TSC1 and TSC2

Employing DHPLC as the primary mutation screening approach, we have studied 660 TSC patients. This set includes 224 patients which we have previously reported[10]. The remainder are a larger new group, not previously reported, for which the analysis is not yet complete. Specifically, not all assays have been performed on each patient, some DNA samples are depleted, and careful review of clinical data has yet to be performed to remove patients who fail to meet TSC diagnostic criteria. Overall we have identified mutations in TSC1 or TSC2 in 469 (71%) patients (Figure 4-1, ***Table 4-1***). This overall set of TSC patients and mutations is much larger than any other previously reported.

The large deletion and duplication mutations in this set have been identified through the use of quantitative PCR (QPCR) and long range PCR (LRPCR) assays. We have developed these methods for use in the analysis of TSC1 and TSC2[10,11]. The QPCR assay is performed by simultaneous amplification of 2-3 amplicons within TSC2 and a control amplicon, all with FAM-labeled primers and of different size, with only 18 cycles of PCR, followed by ABI 3100 capillary electrophoresis. The method is sensitive, quantitative, and quite reproducible in our hands in the detection of reduced (indicating allele loss) or increased (indicating duplication of an allele) representation of an exon of TSC2 in the starting material[10]. The LRPCR assay involves the generation of 19 amplicons providing overlapping coverage of the entire genomic extent of the TSC2 gene. The products vary in size from 1.7 - 11.6 kb and are analyzed by gel electrophoresis[11].

4.7 Analysis of Mutation Distribution and Frequency in TSC1 and TSC2

We observe that there is a very broad spectrum of mutations in both TSC1 and TSC2, both in terms of type and distribution of mutations among the exons of these genes (Figure 4-1). However, the distribution of mutations is not completely uniform. Some exons of each gene are not involved by mutation in this set of patients, including the alternatively spliced exons 25 and 31 of TSC2, but also TSC1 exons 3, 6, 16, 22, and 23; and TSC2 exon 15. TSC2 mutations are much more common than TSC1 mutations (4.2:1 ratio), and this ratio is higher than that predicted by their relative genomic extents and coding regions. This

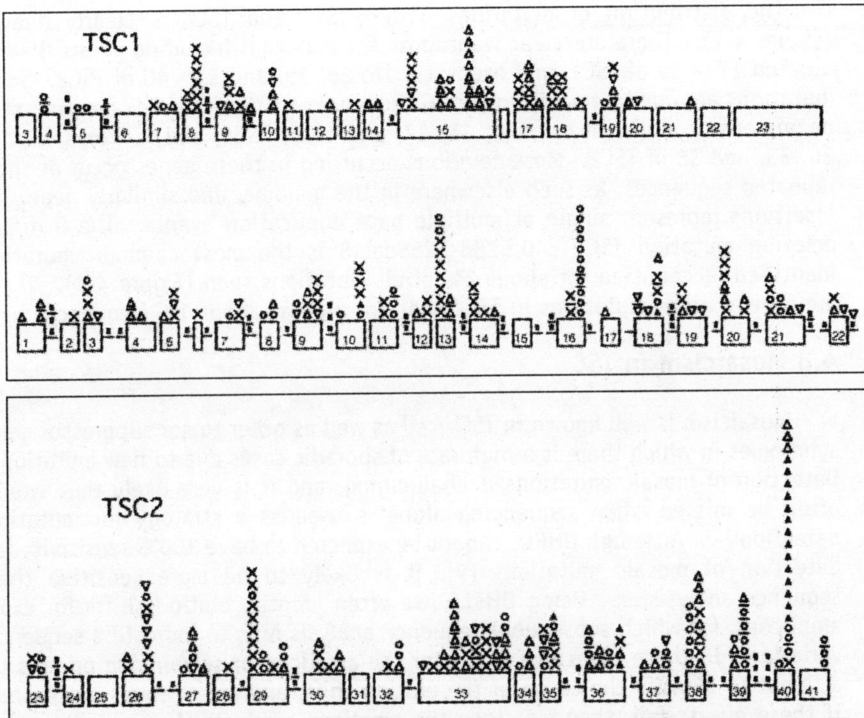

Figure 4-1. Map of the sites of 409 small mutations in TSC1 and TSC2 identified in our laboratory. Proportional drawings (boxes) of all of the exons of TSC1 and TSC2 are shown. Intron regions are not drawn to scale and are expanded only when a mutation is present. Mutation symbols are: x, nonsense; _, splice site; _, deletion; _, insertion; _, missense; _, in-frame deletion. A line separates mutations of the same type occurring at nearby but not identical nucleotide positions.

observation may reflect a higher intrinsic mutation rate in TSC2[4]. Missense, in-frame deletion, and large deletion mutations are very rare in TSC1, although we have identified several possible and one proven TSC1 missense mutations (370T>C 50L>P; 403T>C 61L>P (confirmed by parental analysis); 499T>G 93L>R; 790G>C 190R>P; 958G>C 246R>T). Two TSC1 deletion mutations have also been identified, including one characterized at the molecular level as a deletion of 1279 bp, extending from intron 16 to exon 18. Interestingly, the frequency of small deletion, missense, nonsense, splice, and big deletion mutations are all approximately the same in TSC2 (range 8.2% - 12.4%).

The distribution of mutations within TSC1 and TSC2 is clearly uneven (Figure 4-1). There are clear warmspots for mutation including exons 8 - 10, 15, and 17 - 19 of TSC1, and exons 13, 16, 26, 29, and 33 - 40 of TSC2. Some mutations are due to transitions of CG sequences to TG sequences creating stop or missense codons: in exons 8, 15, 17, and 18 of TSC1; and in exons 14, 16, 20, 33, and 38 of TSC2. Most deletions occurring in these genes occur at short repeated sequences, as seen elsewhere in the genome, and similarly nearly all insertions represent single or multiple base duplication events. The in-frame deletion mutation TSC2 E40:5238-5255del18 is the most common mutation identified, accounting for about 2% of all mutations seen (Figure 4-1). There are several other mutations in TSC2 that are seen at about 1% frequency.

4.8 Mosaicism in TSC

Mosaicism is well known in TSC[12-14] as well as other tumor suppressor gene syndromes in which there is a high rate of sporadic cases due to new mutations. Detection of mosaic mutations is challenging, and it is very likely they would often be missed when sequencing alone is used as a strategy for mutation detection[15]. Although DHPLC cannot be expected to have 100% sensitivity for detection of mosaic mutations [9], it is likely to be more sensitive than sequence analysis[16]. Using DHPLC, we often identify elution shifts for exon amplicons, for which subsequent sequence analysis fails to indicate a sequence variation. In these cases, we may alter the amplicon or sequencing primers to attempt to identify the cause of the elution shift, and are at times successful. If these efforts fail, then we clone the amplicon products (Figure 4-2). Fifty clones are chosen and screened with vector primers to ensure that they all contain the correct sized insert. At least 25 clones and a DNA sample known not to contain a sequence variant (control) are then amplified using the amplicon primers. The clone amplicons are then individually mixed with the control product, heteroduplexed, and analyzed on the DHPLC machine. Clones which give an elution shift in this analysis that is similar to (but often more prominent than) that seen with the original patient DNA sample are then sequenced. The mutation is typically easily identified in this case, as there is a single allele present in the clone. To verify that the sequence variant is real and not the product of PCR or cloning artifact, we require that at least three independent clones show the same sequence variation. We also review the sequence traces of the patient for that amplicon looking for minor evidence of the sequence variant. When performed in this way, in the last 120 patients studied we identified mosaic mutations in 4 (3%) patients. By counting the number of clones containing the mutation, we determined that the level of mosaicism in these patients was 7 - 26 percent (***Table 4-2***).

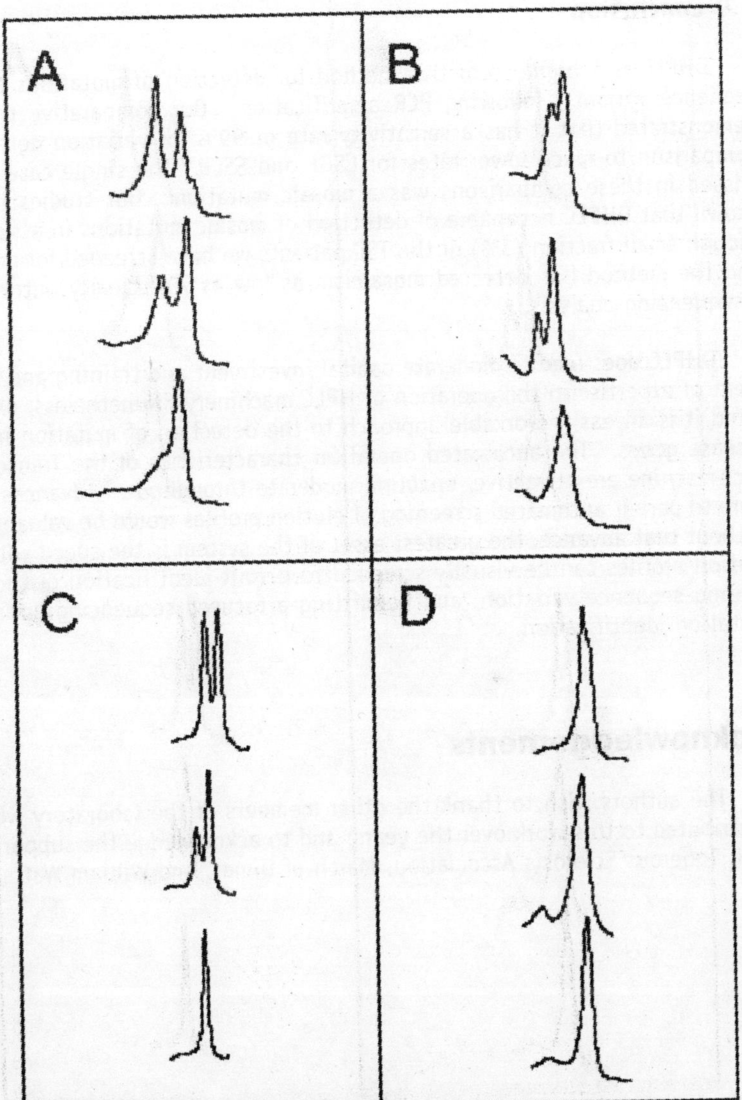

Figure 4-2. Mosaic mutation detection. Each quadrant of this figure shows a compilation of three DHPLC elution profiles. The bottom is a tracing from a control DNA sample, the middle is a tracing from a patient for whom no sequence variation could be identified by direct sequencing of this amplicon, and the top is from a mixing study in which cloned allele amplicons were mixed with a control amplicon. A. TSC2 E38:5034delC. B. TSC2 E18:2095C>T, 699Q>X. C. TSC2 I6:648+1G>A. D. TSC2 E7:671delC.

4.9 Conclusion

DHPLC is a highly sensitive method for detection of mutations and other sequence variants, following PCR amplification. Our comparative trials have demonstrated that it has a sensitivity rate of 99% for variation detection, in comparison to much lower rates for CSGE and SSCP. The single case that was missed in these comparisons was a mosaic mutation. Our studies have also shown that DHPLC is capable of detection of mosaic mutations in a substantial though small fraction (3%) of the TSC patients we have screened for mutations, and the method has detected mosaicism as low as 7%, clearly outperforming resequencing analysis.

DHPLC does require moderate capital investment and training and development of expertise in the operation of HPLC machinery. Nonetheless, once mastered it is an easily workable approach to the detection of mutation in human disease genes. The automated operation characteristics of the Transgenomic, Inc. machine are attractive, enabling moderate throughput. Advances in software to permit automated screening of elution profiles would be valuable. Even without that advance, the greatest asset of the system is the speed with which elution profiles can be visually screened to permit identification of those containing sequence variation, and permitting a focused sequencing approach to mutation identification.

Acknowledgements

The authors wish to thank the other members of the laboratory who have contributed to this work over the years; and to acknowledge the support of the NIH, Tuberous Sclerosis Association, March of Dimes, and William Watts.

Table 4-1: Distribution of mutations in TSC1 and TSC2.

	TSC1		TSC2	
Mutation	#	%	#	%
Insertion	8	1.2	28	4.2
Deletion	28	4.2	60	9.1
In-frame deletions	0		19	2.9
Missense	8	1.2	72	10.9
Nonsense	32	4.8	82	12.4
Splice	14	2.1	54	8.2
Big deletions	2	0.3	61	9.2
Duplications	0		1	0.2
Subtotal	92	14	377	57

Total number of patients analyzed is 660.

Table 4-2: Mosaic mutations found in TSC2 through allele cloning.

Exon/Intron	Mutation	Clone frequency	Estimated percent mosaicism
TSC2 I6	648+1G>A	6/46	26%
TSC2 E18	2095C>T699Q>X	4/32	25%
TSC2 E38	5034delC	1/28*	7%
TSC2 E7	671delC	3/54	11%

*Note that this mutation was confirmed by allele-specific PCR since it was seen in only a single cloned amplicon.

References

1. ***Gomez M, Sampson J, Whittemore V:*** The tuberous sclerosis complex, ed 3. Oxford, *Oxford University Press*, 1999.

2. ***van Slegtenhorst M, de Hoogt R, Hermans C, Nellist M, Janssen B, Verhoef S, Lindhout D, van den Ouweland A, Halley D, Young J, Burley M, Jeremiah S, Woodward K, Nahmias J, Fox M, Ekong R, Osborne J, Wolfe J, Povey S, Snell RG, Cheadle JP, Jones AC, Tachataki M, Ravine D, Kwiatkowski DJ:*** Identification of the tuberous sclerosis gene TSC1 on chromosome 9q34. *Science 1997*;277:805-8.

3. ***The European chromosome 16 tuberous sclerosis consortium:*** Identification and characterization of the tuberous sclerosis gene on chromosome 16. *Cell* 1993;75:1305-1315.

4. ***Cheadle JP, Reeve MP, Sampson JR, Kwiatkowski DJ:*** Molecular genetic advances in tuberous sclerosis. *Hum Genet 2000*;107:97-114.

5. ***Ganguly T, Dhulipala R, Godmilow L, Ganguly A:*** High throughput fluorescence-based conformation-sensitive gel electrophoresis (F-CSGE) identifies six unique BRCA2 mutations and an overall low incidence of BRCA2 mutations in high-risk BRCA1-negative breast cancer families. *Hum Genet 1998*;102:549-56.

6. ***Ravnik-Glavac M, Glavac D, Dean M:*** Sensitivity of single-strand conformation polymorphism and heteroduplex method for mutation detection in the cystic fibrosis gene. *Hum Mol Genet 1994*;3:801-7.

7. ***van Orsouw NJ, Dhanda RK, Rines RD, Smith WM, Sigalas I, Eng C, Vijg J:*** Rapid design of denaturing gradient-based two-dimensional electrophoretic gene mutational scanning tests. *Nucleic Acids Res 1998*;26:2398-406.

8. ***Choy YS, Dabora SL, Hall F, Ramesh V, Niida Y, Franz D, Kasprzyk-Obara J, Reeve MP, Kwiatkowski DJ:*** Superiority of denaturing high performance liquid chromatography over single-stranded conformation and conformation-sensitive gel electrophoresis for mutation detection in TSC2. *Ann Hum Genet 1999*;63:383-91.

9. ***Roberts PS, Jozwiak S, Kwiatkowski DJ, Dabora SL:*** Denaturing high-performance liquid chromatography (DHPLC) is a highly sensitive, semi-automated method for identifying mutations in the TSC1 gene. *J Biochem Biophys Methods 2001*;47:33-37.

10. ***Dabora SL, Jozwiak S, Franz DN, Roberts PS, Nieto A, Chung J, Choy YS, Reeve MP, Thiele E, Egelhoff JC, Kasprzyk-Obara J, Domanska-Pakiela D, Kwiatkowski DJ:*** Mutational Analysis in a Cohort of 224 Tuberous Sclerosis Patients Indicates Increased Severity of TSC2, Compared with TSC1, Disease in Multiple Organs. *Am J Hum Genet 2001*;68:64-80.

11. ***Dabora SL, Nieto AA, Franz D, Jozwiak S, Van Den Ouweland A, Kwiatkowski DJ:*** Characterisation of six large deletions in TSC2 identified using long range PCR suggests diverse mechanisms including Alu mediated recombination. *J Med Genet 2000*;37:877-83.

12. ***Verhoef S, Bakker L, Tempelaars AM, Hesseling-Janssen AL, Mazurczak T, Jozwiak S, Fois A, Bartalini G, Zonnenberg BA, van Essen AJ, Lindhout D, Halley DJ, van den Ouweland AM:*** High rate of mosaicism in tuberous sclerosis complex. *Am J Hum Genet 1999*;64:1632-7.

13. ***Rose VM, Au KS, Pollom G, Roach ES, Prashner HR, Northrup H:*** Germ-Line Mosaicism in Tuberous Sclerosis: How Common? *Am J Hum Genet 1999*;64:986-992.

14. ***Sampson JR, Maheshwar MM, Aspinwall R, Thompson P, Cheadle JP, Ravine D, Roy S, Haan E, Bernstein J, Harris PC:*** Renal cystic disease in tuberous sclerosis: role of the polycystic kidney disease 1 gene. *Am J Hum Genet 1997*;61:843-51.

15. ***Kwiatkowska J, Wigowska-Sowinska J, Napierala D, Slomski R, Kwiatkowski DJ:*** Mosaicism in tuberous sclerosis as a potential cause of the failure of molecular diagnosis. *N Engl J Med 1999*;340:703-7.

16. ***Jones AC, Sampson JR, Cheadle JP:*** Low level mosaicism detectable by DHPLC but not by direct sequencing. *Hum Mutat 2001*;17:233-4.

CHAPTER

5

RAPID ANALYSIS TO DETECT BASE CHANGES IN HUMAN MITOCHONDRIAL DNA BY RESTRICTION ENZYME ANALYSIS FOLLOWED BY DENATURING HIGH PERFORMANCE LIQUID CHROMATOGRAPHY

Patricia E. McAndrew and Michael A. Marino
Applied Genomics & Molecular Genetics Division

Contact:
Dr. Petra Krause
Product Manager Bioconsumables - Kits, Transgenomic, Inc.
12325 Emmett Street, Omaha, NE 68164
E-mail: pkrause@transgenomic.com

9664027-7-4/03/$0.00+$.50 *From:* **Genetic Variance Detection~Nuts & Bolts of DHPLC in Genomics** (pp.93-116)
 Edited by: Karl H. Hecker, Ph.D.

5.1 Introduction

Mitochondria are DNA-containing organelles found within the cytoplasm of eukaryotic cells. Their main function is to provide energy for cellular activity in the form of ATP through the process of oxidative phosphorylation. Sequence analysis of the mitochondrial genome[1] revealed this 16.5-kb circular molecule encodes 37 genes necessary for mitochondrial function. Mitochondrial DNA molecules are maternally inherited and can be present at up to 10,000 copies per cell. It is estimated that the mutation rate in mtDNA is 10-times higher than that for nuclear DNA due to the combined effects of exposure to reactive oxygen species, the lack of protective histones, and the lack of efficient repair mechanisms[2,3].

Molecular analysis of the mitochondrial genome has applications in a wide range of study areas. Since 1988 defects in the human mitochondrial genome have been reported for several degenerative diseases, cancer, and aging[2-7]. Mitochondrial DNA analysis is increasingly being utilized in forensic cases where only limited amounts of degraded material are available[8]. In addition, the study of mitochondrial DNA has also provided insights into human evolution and genetic population studies[9].

Several approaches have been undertaken to analyze the mitochondrial genome including direct sequence analysis, SSCP, TGGE, and DHPLC[8-12]. An efficient method to rapidly screen mtDNA for base changes requires an extremely sensitive system capable of detecting low levels of heteroplasmy. This is the coexistence of normal and altered/mutant mtDNA molecules within an individual and can be maternally inherited or arise through somatic mutation. For many degenerative diseases, the proportion and distribution of this heteroplasmy often determines the severity of clinical symptoms[2,3]. DHPLC has been shown by several researchers to be a highly sensitive method for detecting low levels of heteroplasmy[13] and somatic mutations in cases where other methods failed[11]. An efficient assay is necessary because disorders resulting from mtDNA mutations often present with a wide range of clinical symptoms and have complex inheritance patterns. The recent identification of mutations in nuclear genes that cause disease by introducing mutation in mtDNA further complicates the molecular diagnosis of mitochondrial disorders[14,15]. We present a rapid, sensitive DHPLC method to screen the human mitochondrial genome for base changes. Results from this method will contribute to the growing understanding of the complexities of mitochondrial pathology and can be used for a variety of different applications involving mtDNA[1].

5.2 Materials and Methods

5.2.1 Amplification of Mitochondrial DNA

Primer sets were designed to amplify the entire human mitochondrial genome in eighteen overlapping segments. The mitochondrial regions amplified, and the size of the polymerase chain reaction (PCR) products are listed in ***Table 5-1***. Mitochondrial base pair designations are consistent with the MITOMAP sequence version. Genomic DNA from cell lines CHR and 9947A were obtained from the National Institutes of Standards and Technologies (NIST; SRM 2392), and genomic DNA from the K562 cell line was purchased from Promega (Madison, WI). The mitochondrial fragments were amplified using the Expand' Long Template PCR System (Roche Molecular Biochemicals, Indianapolis, IN) in 1X PCR buffer 1 with 1.75 mM $MgCl_2$, 350 µM each dNTP (Applied Biosystems, Foster City, CA), 300 nM forward primer, 300 nM reverse primer, 1U Expand' Taq polymerase mix, and 50 - 100 ng of genomic DNA according to the manufacturer's instructions. The components were denatured at 93°C for 2 minutes, followed by 30 cycles of 93°C for 10 seconds; 58°C for 30 seconds; 68°C for 1.5 minutes, with a final extension step of 68°C for 7 minutes. (*See Footnote 1*)

5.2.2 Restriction Enzyme Digestion

Restriction endonucleases were selected that cleave the mtDNA PCR products into ~ 100- to 600-bp fragments (New England Biolabs, Beverly, MA). The restriction enzymes, recommended 10X buffers, fragment sizes, and the corresponding mtDNA regions are listed in ***Table 5-2***. PCR products amplified in mitochondrial sets 1, 2, 3, and 15 do not require a restriction enzyme digestion, and can be set-aside until the next step in the procedure. Each restriction enzyme digestion was performed in a final volume of 100 µl with 88.5 µl of mtDNA PCR product, 10µl of the recommended 10X restriction enzyme buffer, and 1.5µl of the appropriate enzyme(s). The reactions were incubated at 37°C for a minimum of 2 hours, and 9µl were analyzed on the Transgenomic WAVE® DNA Fragment Analysis System at 50°C to check for complete digestion of the PCR products. A 10-µl aliquot of the pUC18/HaeIII digest (Transgenomic, Inc., Omaha, NE) was run as a size standard.

---

[1] This application has been further developed and improved by using a proof reading Polymerase (Optimase® Polymerase, Transgenomic, Inc.) for easier mutation calling. Furthermore the PCR fragment MT #13 has been split into two fragments in order to obtain a more robust application for different qualities of DNA. This resulted in a total number of 19 PCR fragments.

The development has lead to the MitoScreen™ Assay Kit, a kit for Rapid Mutation Screening of the Entire Mitochondrial Genome using Multiplex DHPLC, which is now available from Transgenomic Inc.

Further information can be obtained via www.transgenomic.com or by e-mailing info@transgenomic.com

5.2.3 Heteroduplex Detection by Denaturing High Performance Liquid Chromatography (DHPLC)

The WAVE® System has a DNASep® column with a stationary phase consisting of 2 µm nonporous, alkylated poly(styrene-divinylbenzene) particles, a UV detector set at 260 nm, and two 96-well autosamplers. The mobile phase consists of Solvent A: 0.1M triethylammonium acetate (Transgenomic, Inc.) and Solvent B: 0.1M triethylammonium acetate, 25% (v/v) acetonitrile (Transgenomic, Inc.). The gradient for elution of the restriction enzyme fragments is shown in ***Table 5-3***.

Digested mtDNA PCR products and MT sets 1, 2, 3, and 15 were analyzed by DHPLC using the same gradient conditions as shown in Table 5-3 for the restriction enzyme fragments. Products from each cell line were analyzed individually, and an equal amount of product from each cell line was mixed together and analyzed by DHPLC to detect the presence of single nucleotide polymorphisms (SNPs). The samples were heated to 95°C for 5 minutes and cooled slowly to room temperature (–0.1°C/sec) to allow the formation of heteroduplex molecules. The column temperatures for each fragment were predicted by Wavemaker™ Software version 4.0 (Transgenomic, Inc.) with minor adjustments (***Table 5-4***). An injection volume of 12µl was sufficient for each sample at each temperature, but this may need to be increased if the PCR product yield is low.

5.3 Results

5.3.1 Restriction Enzyme Analysis

Mitochondrial PCR fragments amplified from CHR and 9947A cell lines were digested with the appropriate restriction enzymes and run on the WAVE® System at 50°C. DNA fragments are separated on the basis of size at this non-denaturing temperature. This step is critical to ensure the expected number of peaks is produced and that no partial digestion products are present. All PCR products and restriction enzyme fragments from the CHR and 9947A cell lines had the predicted number of peaks with elution times consistent with their expected sizes. Figure 5-1 shows the products of an AluI restriction enzyme digestion of MT set 18 amplified from K562, CHR, and 9947A cell lines. This example illustrates the potential problems that can result when a restriction fragment length polymorphism (RFLP) is present in one of the samples. The expected peaks with elution times of 11.5, 14, and 15 minutes were present in the CHR and 9947A cell lines. However, the peak with the elution time of 14 minutes is absent in the K562 cell line (shown as an arrow with a dotted line in Figure 5-1) and two additional peaks are present with elution times of 10 and 10.5 minutes (shown as arrows with solid lines). These results indicate that an additional recognition site for AluI is present in the K562 mitochondrial DNA sequence; therefore this cell line should not be used as a control for MT set 18.

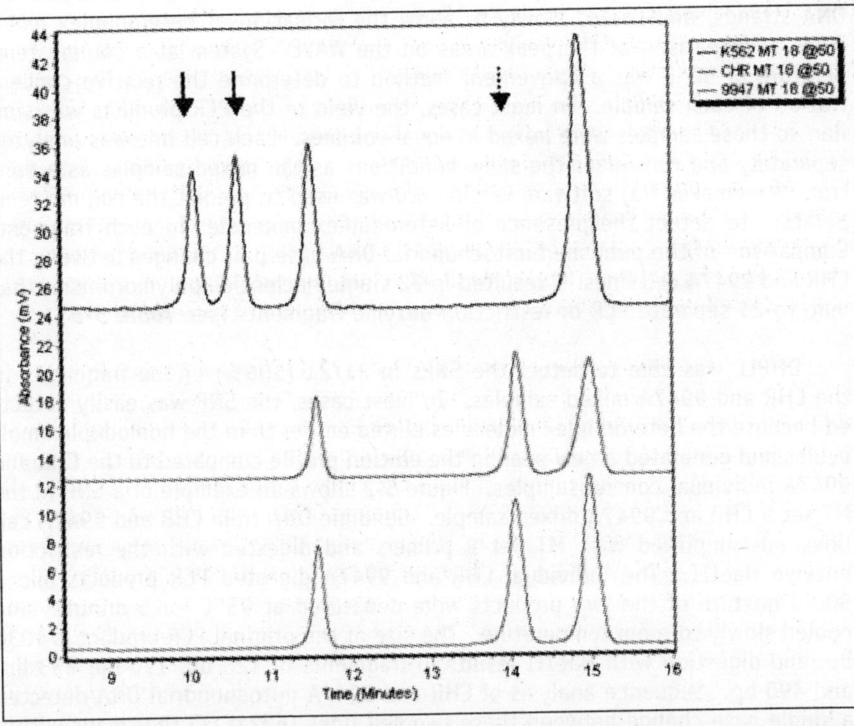

Figure 5-1: Example of Restriction Fragment Length Polymorphism (RFLP) in MT Set 18. Genomic DNA from different cell lines was amplified with MT Set 18 primers and digested with the restriction enzyme AluI at 37°C for 2 hours. The products were analyzed by DHPLC at a column temperature of 50°C. Cell lines CHR and 9947A show the expected AluI fragments of 218 bp, 352 bp, and 458 bp. The restriction enzyme pattern of the K562 cell line is missing the 352 bp fragment (arrow with dashed line), but 2 smaller bands ~ 165 bp and 180 bp are present (arrows with solid line). This pattern indicates that the K562 mitochondrial DNA sequence contains an additional AluI site within the 352-bp fragment. This example demonstrates the importance of analyzing all mitochondrial fragments by DHPLC at 50°C to check that the predicted number of bands with the expected retention times is generated to avoid misinterpretation of the elution profiles under partially denaturing conditions.

5.3.2 Heteroduplex Detection by DHPLC

An equal amount of PCR product or restriction enzyme digested material from the CHR and 9947A cell lines was mixed, heated to 95°C to denature the DNA strands, and cooled slowly to allow the formation of heteroduplex molecules. Comparison of the peak areas on the WAVE® System at a column temperature of 50°C was a convenient method to determine the relative concentration of each sample. In most cases, the yield of the PCR products was similar, so those samples were mixed in equal volumes. Each cell line was analyzed separately and run under the same conditions as the mixed samples as a control. Wavemaker(tm) software version 4.0 was used to predict the column temperature to detect the presence of heteroduplex molecules in each fragment. Comparison of the published mitochondrial DNA base pair changes between the CHR and 9947A cell lines[16] resulted in 32 single nucleotide polymorphisms that map to 21 separate PCR or restriction enzyme fragments (see ***Table 5-5***).

DHPLC was able to detect the SNPs in 21/21 (100%) of the fragments in the CHR and 9947A mixed samples. In most cases, the SNP was easily detected because the heteroduplex molecules eluted earlier than the homoduplex molecules and generated a new peak in the elution profile compared to the CHR and 9947A individual control samples. Figure 5-2 shows an example of a SNP in the MT set 9 CHR and 9947A mixed sample. Genomic DNA from CHR and 9947A cell lines was amplified with MT Set 9 primers and digested with the restriction enzyme HaeIII. The individual CHR and 9947A digested PCR products and a 50:50 mixture of the two products were denatured at 95°C for 5 minutes and cooled slowly to room temperature. The size of the original PCR product is 1036 bp, and digestion with HaeIII results in fragments of 123 bp, 190 bp, 233 bp, and 490 bp. Sequence analysis of CHR and 9947A mitochondrial DNA detected a single base change between these two cell lines (6221C/T) that maps within the 233 bp fragment eluting at 12 minutes (indicated by an arrow in Figure 5-2). The CHR and 9947A mixed sample clearly shows the formation of a heteroduplex peak at a column temperature of 58°C compared to the single 233-bp peak in the individual samples. This example demonstrates the sensitivity of DHPLC to detect this SNP, and interpretation of the result is straightforward based on heteroduplex detection by peak shape.

Since the multiple fragments analyzed in the mitochondrial sets are derived from a restriction enzyme digestion, the relative ratio of the peak heights within a sample can be used to detect the presence of SNPs by DHPLC. Figures 5-3A and 5-3B show two examples in which the peak height was used in combination with the peak shape to detect SNPs in the mixed samples. The individual CHR and 9947A digested PCR products and a 50:50 mixture of the two products were denatured at 95°C for 5 minutes and cooled slowly to room tem-

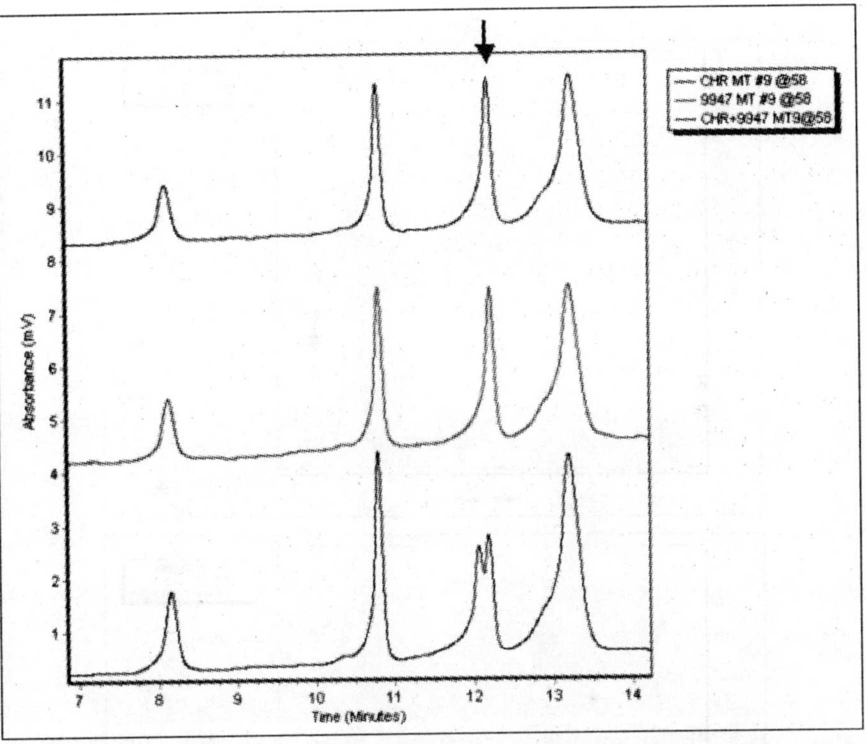

Figure 5-2: Chromatogram of MT Set 9 (58°C). Genomic DNA from CHR and 9947A cell lines was amplified with MT Set 9 primers and digested with the restriction enzyme HaeIII at 37°C for 2 hours. The individual CHR and 9947A digested PCR products and a 50:50 mixture of the two products were denatured at 95°C for 5 minutes and cooled slowly to room temperature. DHPLC analysis was performed at a column temperature of 58°C. The size of the original PCR product is 1036 bp, and digestion with HaeIII results in fragments of 123 bp + 190 bp + 233 bp + 490 bp. Sequence analysis of CHR and 9947A mitochondrial DNA detected a single base change between these two cell lines (6221C/T), which maps within the 233-bp fragment (indicated by arrow). The CHR and 9947A mixed sample clearly shows the formation of a heteroduplex peak compared to the single 233-bp peak in the individual samples. This example demonstrates the sensitivity of the mitochondrial genome scan to detect the single nucleotide polymorphism (SNP) between these two samples. Interpretation of the results is straightforward and based on heteroduplex detection by peak shape.

perature. DHPLC analysis was performed at a column temperature of 56°C for MT Set 8 and 59°C for MT Set 12. Sequence analysis of the mitochondrial DNA detected base changes between these two cell lines (5186G/A in the 396-bp

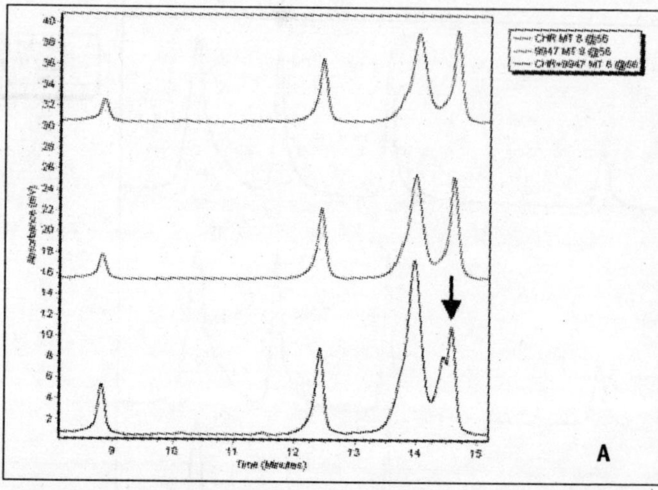

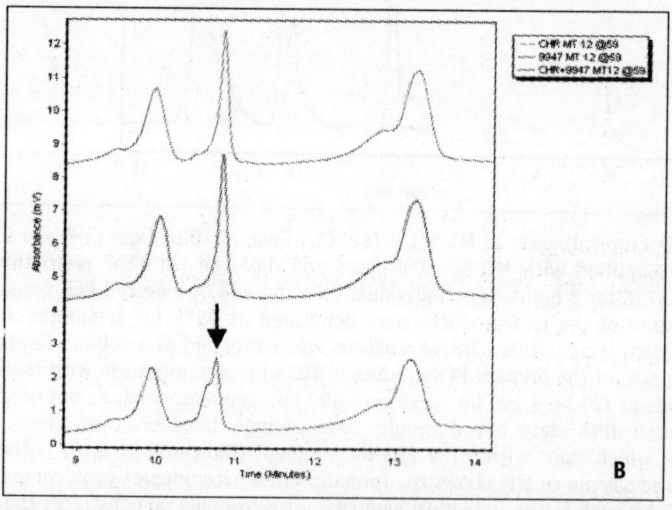

Figure 5-3A-B: Genomic DNA from CHR and 9947A cell lines was amplified with MT Set 8 or MT Set 12 primers and digested with the appropriate restriction enzyme at 37°C for 2 hours. The individual CHR and 9947A digested PCR products and a 50:50 mixture of the two products were denatured at 95°C for 5 minutes and cooled slowly to room temperature. The chromatograms show the DHPLC analysis at a column temperature of 56°C for MT set 8 and 59°C for MT set 12. Sequence analysis of CHR and 9947A mitochondrial DNA detected base changes between these two cell lines (5186G/A - 396-bp fragment MT set 8; 9315T/C - 188-bp fragment MT set 12). These examples demonstrate that although heteroduplex peaks can be identified in the CHR and 9947A mixed samples containing the SNPs (indicated by arrows), interpretation of the results is most sensitive when both peak shape and peak height (mV intensity) are considered.

fragment of MT set 8; 9315T/C in the 188-bp fragment of MT set 12). Comparison of the peak heights in Figure 5-3A reveals the mV intensities of the peaks eluting at 14 and 14.6 minutes are similar in the CHR and 9947A cell lines. However, the relative ratio of the peak heights in the CHR and 9947A mixed sample are different from the individual samples. The peak that elutes at 14.6 minutes has approximately half the mV intensity as the peak that elutes at 14 minutes. These examples demonstrate that although heteroduplex peaks can be identified in the CHR and 9947A mixed samples containing these SNPs, interpretation of the results is most sensitive when both peak shape and peak height (mV intensity) are evaluated.

Figure 5-4 shows the chromatograms of MT set 10 at column temperatures from 56 - 59°C. Heteroduplex peaks can be clearly identified in each CHR and 9947A mixed sample containing a SNP, but interpretation of the results is more complex. The size of the MT set 10 PCR product is 1390 bp, and digestion with MspI results in fragments of 117 bp, 162 bp, 252 bp, 354 bp, and 505 bp. Sequence analysis of CHR and 9947A mitochondrial DNA detected base changes between these two cell lines (6371T/C - 252-bp fragment or peak 3; 6791G/A - 162-bp fragment or peak 2; 7028T/C - 354-bp fragment or peak 4; 7645T/C - 505-bp fragment or peak 5). The arrow in each chromatogram indicates the optimal screening temperature recommended for each fragment. Analysis of the elution profiles is more complex because the 505-bp fragment (peak 5) has a lower melting temperature than the 354bp fragment (peak 4), thus peak 5 elutes before peak 4 at 58°C and 59°C.

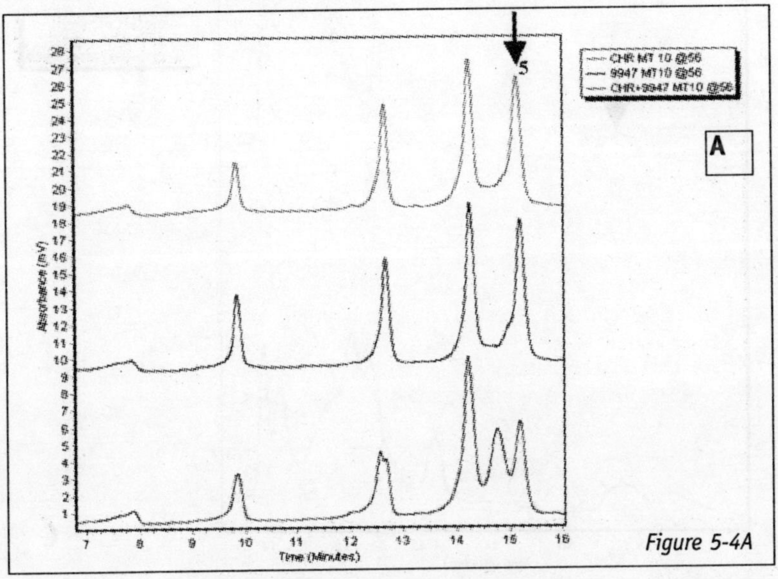

Figure 5-4A

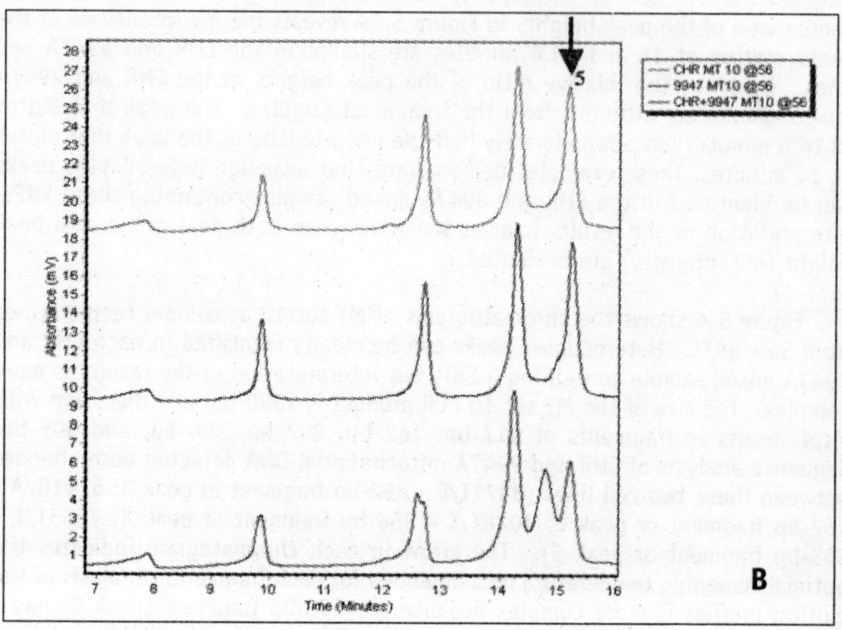
CHR MT 10 @56
9947 MT10 @56
CHR+9947 MT10 @56
5
Absorbance (mV)
Time (Minutes)

B

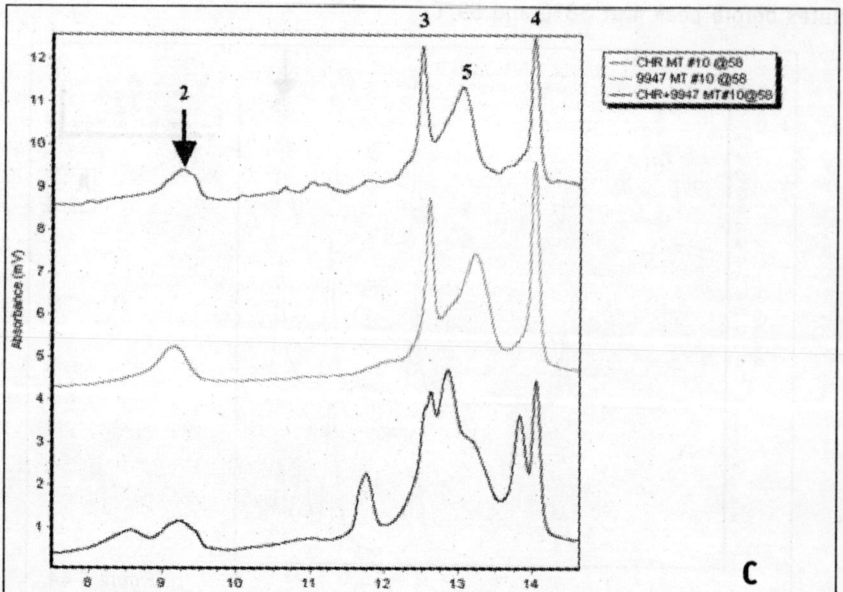
CHR MT #10 @58
9947 MT #10 @58
CHR+9947 MT#10@58
2
3
4
5
Absorbance (mV)
Time (Minutes)

C

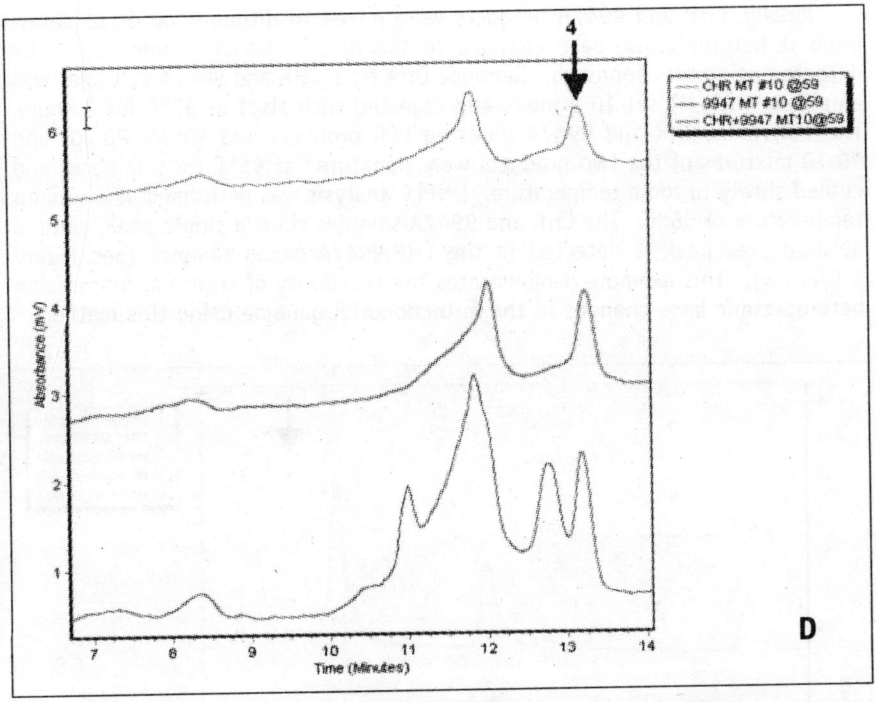

Figure 5-4A-D: Genomic DNA from CHR and 9947A cell lines was amplified with MT set 10 primers and digested with the restriction enzyme MspI at 37°C for 2 hours. The individual CHR and 9947A digested PCR products and a 50:50 mixture of the two products were denatured at 95°C for 5 minutes and cooled slowly to room temperature. DHPLC analysis was performed at column temperatures of 56 - 59°C. The size of the MT Set 10 PCR product is 1390 bp, and digestion with MspI results in fragments of 117 bp, 162 bp, 252 bp, 354 bp, and 505 bp. Sequence analysis of CHR and 9947A mitochondrial DNA detected base changes between these two cell lines (6371T/C - 252-bp fragment or peak 3; 6791G/A - 162-bp fragment or peak 2; 7028T/C - 354-bp fragment or peak 4; 7645T/C - 505-bp fragment or peak 5). The arrow in each chromatogram indicates the optimal screening temperature recommended for each fragment. A heteroduplex peak can be clearly identified in each CHR and 9947A mixed sample containing the SNP, but interpretation of the results is more complex. For example, the 505-bp fragment (peak 5) has a lower melting temperature than the 354-bp fragment (peak 4), thus peak 5 elutes before peak 4 at 58°C and 59°C.

Finally, CHR and 9947A products were mixed in different ratios to determine if heteroplasmic base changes in the mitochondrial genome could be detected using this approach. Genomic DNA from CHR and 9947A cell lines was amplified with MT set 10 primers and digested with MspI at 37°C for 2 hours. The individual CHR and 9947A digested PCR products and 50:50, 80:20, and 90:10 mixtures of the two products were denatured at 95°C for 5 minutes and cooled slowly to room temperature. DHPLC analysis was performed at a column temperature of 56°C. The CHR and 9947A samples show a single peak, while a heteroduplex peak is detected in the CHR:9947A-mixed samples (See Figure 5-5 arrow). This example demonstrates the feasibility of scanning for variable heteroplasmic base changes in the mitochondrial genome using this method.

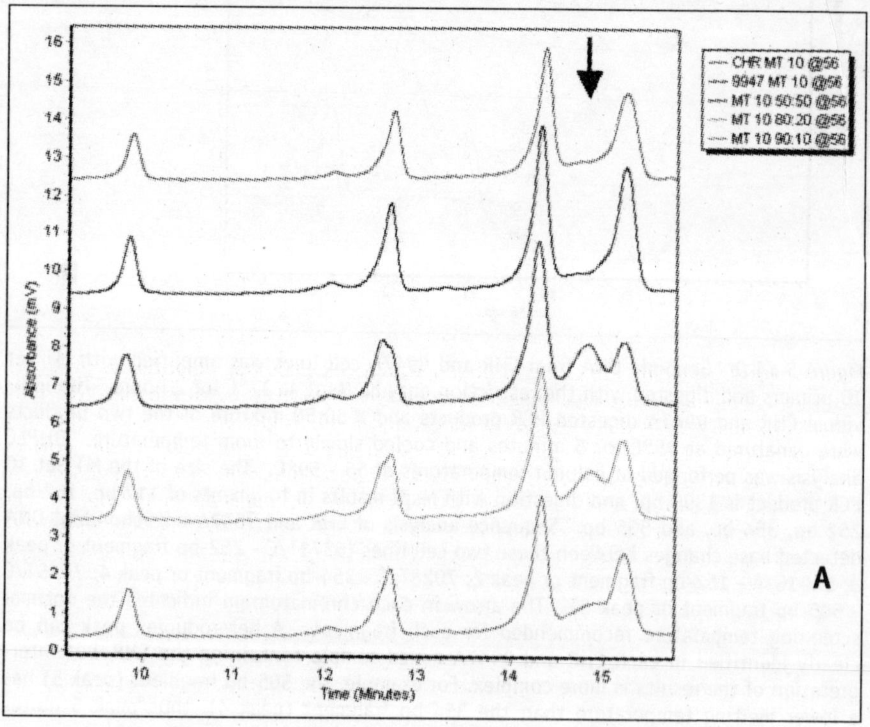

Figure 5-5A

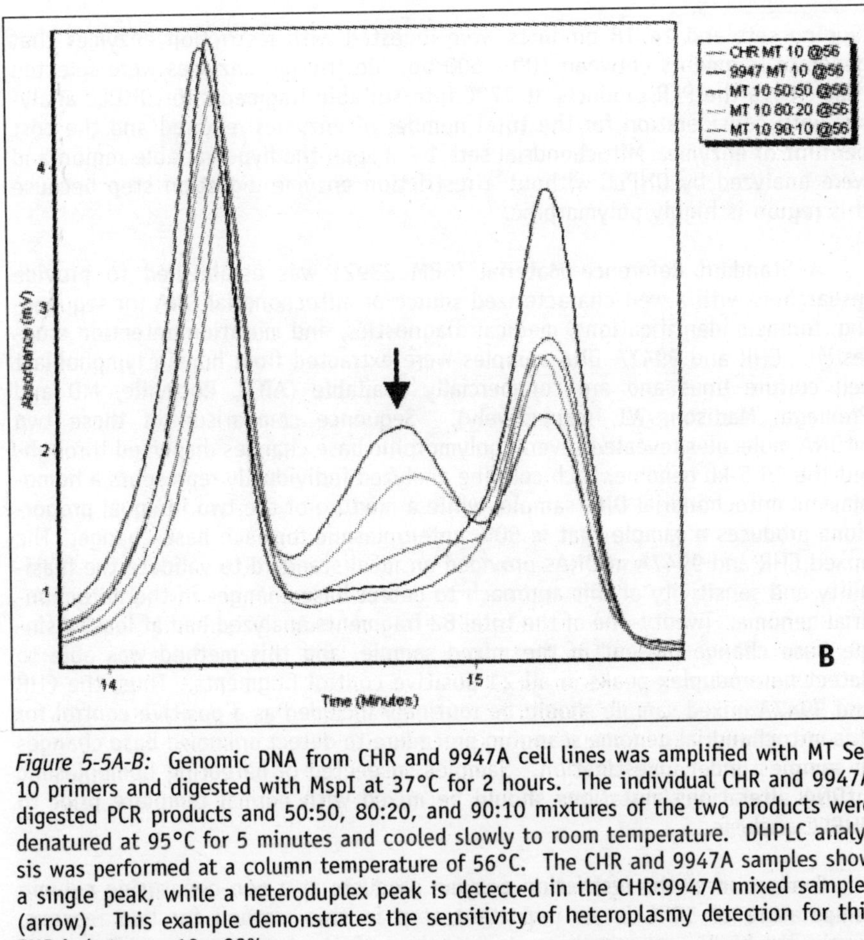

Figure 5-5A-B: Genomic DNA from CHR and 9947A cell lines was amplified with MT Set 10 primers and digested with MspI at 37°C for 2 hours. The individual CHR and 9947A digested PCR products and 50:50, 80:20, and 90:10 mixtures of the two products were denatured at 95°C for 5 minutes and cooled slowly to room temperature. DHPLC analysis was performed at a column temperature of 56°C. The CHR and 9947A samples show a single peak, while a heteroduplex peak is detected in the CHR:9947A mixed samples (arrow). This example demonstrates the sensitivity of heteroplasmy detection for this SNP is between 10 - 20%.

5.4. Discussion

Molecular studies of the mitochondrial genome are of interest to members of the medical community determining links between mtDNA base changes and disease, the forensics community for identification purposes, and individuals studying human evolution[8,9,17]. The goal of this study was to develop a protocol to efficiently screen the 16.5-kb mitochondrial genome for mutations and polymorphisms by DHPLC. The mitochondrial genome was amplified in 18 overlapping sets and 14/18 products were digested with restriction enzymes that generate fragments between 100 - 600 bp. Restriction enzymes were selected that cleave the PCR products at 37°C into suitable fragments for DHPLC analysis, with consideration for the total number of enzymes required and the cost per unit of enzyme. Mitochondrial sets 1 – 3 span the hypervariable region and were analyzed by DHPLC without a restriction enzyme digestion step because this region is highly polymorphic.

A Standard Reference Material (SRM 2392) was established to provide researchers with a well-characterized source of mitochondrial DNA for sequencing, forensic identifications, medical diagnostics, and mutation detection studies[16]. CHR and 9947A DNA samples were extracted from human lymphoblast cell culture lines and are commercially available (ATCC, Rockville, MD and Promega, Madison, WI, respectively). Sequence comparison of these two mtDNA molecules revealed several polymorphic base changes dispersed throughout the 16.5-kb genome. Each cell line analyzed individually represents a homoplasmic mitochondrial DNA sample, while a mixture of the two in equal proportions produces a sample that is 50% heteroplasmic for each base change. The mixed CHR and 9947A mtDNAs provided an ideal standard to validate the feasibility and sensitivity of this approach to detect base changes in the mitochondrial genome. Twenty-one of the total 62 fragments analyzed had at least a single base change present in the mixed sample, and this method was able to detect heteroduplex peaks in all 21 positive control fragments. Thus, the CHR and 9947A mixed sample should be routinely included as a positive control for this mitochondrial genome scanning procedure to detect unknown base changes in samples under investigation. Samples suspected of harboring homoplasmic mtDNA alterations/mutations should be mixed with normal template prior to DHPLC analysis.

Evaluation of the restriction enzyme products at a non-denaturing column temperature of 50°C is an important step in this method for two reasons. Firstly, the highly polymorphic nature of the mtDNA sequence can result in base changes that cause the gain or loss of an enzyme recognition site. Figure 5-1 illustrates an example of the gain of an AluI site in MT set 18 in the K562 cell line. It would be virtually impossible to correctly interpret the elution profile of this sample at elevated column temperatures. Secondly, this step is important to determine that no partial restriction enzyme products are present that would also lead to misinterpretation of the results.

One clear advantage of this technique is that once a heteroduplex peak is detected, the corresponding region of the mitochondrial genome can be immediately located. For example, a heteroduplex peak was detected in the mixed MT set 9 CHR and 9947A sample (Figure 5-2; elution time 12 minutes). Table

5-2 indicates that this 233-bp peak corresponds to mitochondrial base pairs 6029-6261. The mitochondrial genome scan can be used to rapidly identify regions in mtDNA that contain putative base changes, so only a limited number of sequencing reactions are needed to confirm the DHPLC results. This technique does require careful examination of the elution order of the peaks at different temperatures to ensure the proper peak is identified for further analysis, since each fragment has different melting characteristics. The elution profiles can be complex for some fragments (Figure 5-4C-D), but the CHR and 9947A positive control fragments provide a well-characterized reference to guide the identification of specific peaks, and thus the location in the mitochondrial genome for further analysis.

This report describes a rapid, highly sensitive DHPLC method to scan for genetic alterations in the human mitochondrial genome. Commercially available cell lines were used to validate the accuracy of this procedure to detect polymorphic changes dispersed throughout the 16.5-kb sequence. This approach can be used to detect the presence of both inherited or somatic mtDNA base changes. Identification and characterization of these base changes is an important first step in the challenging task of assigning functional consequences to alterations in the mitochondrial genome.

Table 5-1: Mitochondrial DNA Fragments for Whole Genome Scan.

MT Set	MT Region Amplified	Size (bp)
MT#1	15975-16410 bp	436
MT#2	16342-102 bp	330
MT#3	29-480 bp	452
MT#4	368-1713 bp	1346
MT#5	1650-2841 bp	1192
MT#6	2415-3812 bp	1398
MT#7	3430-4429 bp	1000
MT#8	4181-5489 bp	1309
MT#9	5348-6383 bp	1036
MT#10	6319-7708 bp	1390
MT#11	7645-8785 bp	1141
MT#12	8644-9459 bp	816
MT#13	398-11398 bp	2001
MT#14	11323-12853 bp	1531
MT#15	12754-13265 bp	512
MT#16	13173-14611 bp	1439
MT#17	14428-15591 bp	1164
MT#18	15425-16452 bp	1028

Table 5-2: Restriction Enzyme Digestion of Mitochondrial DNA PCR Products.

Mitochondrial Set	Restriction Enzyme	10X NEB Buffer	Fragment Sizes	Mitochondrial Region (bp)
MT #1	-	-	436	15975-16410
MT #2	-	-	331	16342-102
MT #3	-	-	452	29-480
MT #4	MboI	NEB-3	211	740-950
			276	951-1226
			372	368-739
			487	1227-1713
MT #5	HaeIII	NEB-2	273	2569-2841
			394	2175-2568
			525	1650-2174

Table 5-2: Restriction Enzyme Digestion of Mitochondrial DNA PCR Products. *(continued)*

Mitochondrial Set	Restriction Enzyme	10X NEB Buffer	Fragment Sizes	Mitochondrial Region (bp)
MT #6	DdeI	NEB-3	125	3067-3192
			210	2857-3066
			279	3535-3812
			342	3193-3534
			442	2415-2856
MT #7	HaeIII	NEB-2	109	3851-3959
			178	3430-3608
			242	3609-3850
			471	3960-4429
MT #8	MspI	NEB-2	135	4712-4846
			248	5243-5489
			396	4847-5242
			530	4181-4711
MT #9	HaeIII	NEB-2	123	6262-6383
			190	5839-6028
			233	6029-6261
			490	5348-5838
MT #10	MspI	NEB-2	117	6572-6688
			162	6689-6850
			252	6319-6571
			354	6851-7204
			505	7205-7708
MT #11	HaeIII	NEB-2	141	8252-8392
			181	8393-8573
			213	8574-8785
			242	7645-7887
			364	7888-8251
MT #12	DdeI	NEB-3	188	9273-9459
			238	8644-8882
			390	8883-9272
MT #13	AluI	NEB-2	247	9398-9645
			312	10600-10911
			366	10234-10599
			440	10912-11398
			588	9646-10233
MT #14	HaeIII + MspI	NEB-2	178	12124-12301
			365	11323-11688
			435	11689-12123
			553	12302-12853
MT #15	-	-	512	12754-13265

Table 5-2: Restriction Enzyme Digestion of Mitochondrial DNA PCR Products. *(continued)*

Mitochondrial Set	Restriction Enzyme	10X NEB Buffer	Fragment Sizes	Mitochondrial Region (bp)
MT #16	AluI + DdeI	NEB-2	129	14306-14434
			174	14435-14611
			289	14017-14305
			381	13173-13554
			462	13555-14016
MT #17	MboI	NEB-3	191	14869-15059
			236	15357-15591
			297	15060-15356
			440	14428-14868
MT #18	AluI	NEB-2	218	15778-15995
			352	15425-15777
			458	15996-16452

Table 5-3: WAVE System Analysis Gradient

Step	Time	%A	%B
Loading	0.0	65	35
Start Gradient*	1.0	60	40
Stop Gradient*	17.0	28	72
Start Clean	17.1	60	40
Stop Clean	17.2	60	40
Start Equilibrate	17.3	60	40
Stop Equilibrate	17.4	60	40

* The results included in this protocol were obtained using the gradient listed above. Further studies indicated the analysis time could be reduced using the adjusted gradient of 45 - 67 % B from 0.5 to 11.5 minutes. This gradient has the same slope as the original gradient, therefore the resolution of the peaks is not altered.

Table 5-4: Screening Temperatures for Heteroduplex Detection in Mitochondrial Fragments.

MT Set	Fragment Size	Oven Temperature (°C)
MT #1	436	58
MT #2	331	60
MT #3	452	57, 59
MT #4	211	57-59
	276	57, 58
	372	57
	487	57, 58
MT #5	273	59
	394	56
	525	56
MT #6	125	57
	210	58
	279	59
	342	59
	442	59
MT #7	109	60
	178	58, 59
	242	57, 58
	471	57
MT #8	135	56
	248	57
	396	56, 57
	530	55
MT #9	123	59
	190	59
	233	58
	490	58
MT #10	117	55
	162	57, 58
	252	57, 58
	354	57-59
	505	56

Table 5-4: Screening Temperatures for Heteroduplex Detection in Mitochondrial Fragments. *(continued)*

MT Set	Fragment Size	Oven Temperature (°C)
MT #11	141	56
	181	56
	213	56
	242	56
	364	59
MT #12	188	59
	238	57
	390	57
MT #13	247	57, 58
	312	54, 55
	366	54, 55
	440	57
	588	54, 55
MT #14	178	56
	365	56
	435	58
	553	54
MT #15	512	59
MT #16	129	59
	174	55, 56
	289	55-57
	381	57
	462	55-57
MT #17	191	59
	236	57, 58
	297	57-59
	440	56, 57
MT #18	218	55
	352	57
	458	57

Table 5-5: Mitochondial DNA Sequence Comparison of CHR and 9947A Cell Lines.

MT Set	Location in MITOMAP	Restriction Fragment Size (bp)	CHR	9947A
2	73	331	G	A
	93	452	A	G
	204	452	C	T
3	207	452	A	G
	214	452	A	G
	309.2	452	-	ins(C)
4	709	372	A	G
5	1719	525	A	G
	2706	273	G	A
7	4135	471	T	C
8	5186	396	G	A
9	6221	233	C	T
10	6371	252	T	C
	6791	162	G	A
	7028	354	T	C
	7645	505	T	C
11	7861	242	T	C
	8448	181	T	C
	8503	181	C	T
12	9315	188	T	C
14	11719	435	A	G
	11878	435	C	T
	12612	553	G	A
	12705	552	T	C
16	13572	462	T	C
	13708	462	A	G
	13759	462	G	A
	13966	462	G	A
	14470	174	C	T
17	14767	440	T	C
18	16183	458	C	A
	16189	458	C	T

References

1. ***Anderson S, Bankier AT, Barrell BG, de Bruijn MH, Coulson AR, Drouin J, Eperon IC, Nierlich DP, Roe BA, Sanger F, Schreier PH, Smith AJ, Staden R, Young IG:*** (1981) Sequencing and organization of the human mitochondrial genome. *Nature* 290, 457-465.

2. ***Lightowlers RN, Chinnery PJ, Turnbull GM, Howell N:*** (1997) Mammalian mitochondrial genetics: heredity, heteroplasmy and disease. *Trends in Genet.* 13, 450-455.

3. ***Wallace, DC:*** (1999) Mitochondrial Diseases in Man and Mouse. *Science*, 283, 1482-1488.

4. ***Fliss MS, Usadel H, Caballero OT, Wu L, Buta MR, Eleff SM, Jen, Sidransky D:*** (2000) Facile detection of mitochondrial DNA mutations in tumors and bodily fluids. *Science* 287, 2017-2019.

5. ***Lin MT, Simon DK, Ahn CH, Kim LH, Beal MF:*** (2002) High aggregate burden of somatic mtDNA point mutations in aging and Alzheimer's disease brain. *Hum Mol Genet* 11, 133-45.

6. ***Polyak K, Li Y, Shu H, Kengauer C, Willson JKV, Markowitz SD, Trush MA, Kinzler KS, Vogelstein B:*** (1998) Somatic mutations of the mitochondrial genome in human colorectal tumours. *Nat Genet* 20, 291-293.

7. ***Yeh JJ, Lunetta KL, van Orsouw NJ, Moore FD Jr, Mutter GL, Vijg J, Kahia PLM, Eng C:*** (1999) Somatic mitochondrial DNA (mtDNA) mutations in papillary thyroid carcinomas and differential mtDNA sequence variants in cases with thyroid tumors. *Oncogene* 19, 2060-6066.

8. ***Salas A, Rasmussen EM, Lareu MV, Morling N, Carracedo A:*** (2001) Fluroescent SSCP of overlapping fragments (FSSCP-OF): a highly sensitive method for the screening of mitochondrial DNA variation. *Forensic Sci Int* 124, 97-103.

9. ***Ingman M, Gyllensten U:*** (2001) Analysis of the complete human mtDNA genome: methodology and inferences for human evolution. *J Hered*, 92, 454-61.

10. ***Lui MR, Pan KF, Li ZF, Wang Y, Deng JH, Zhang L, Lu YY:*** (2002) Rapid screening mitochondrial DNA mutation by using denaturing high-performance liquid chromatography. *World J Gatroenterol*, 8, 426-430.

11. ***van den Bosch BJC, de Coo RFM, Scholte HR, Nijland JG, van den Bogaard R, de Visser M, de Die-Smulders CEM, Smeets HJM:*** (2000) Mutation analysis of the entire mitochondrial genome using denaturing high performance liquid chromatography. *Nucleic Acids Research*, Vol. 28, e89.

12. ***Tan DK, Bai RK, Wong LJ:*** (2002) Comprehensive scanning of somatic mitochondrial DNA mutations in breast cancer. *Cancer Res* 15, 972-976.

13. ***Jones AC, Sampson JR, Cheadle JP:*** (2001) Low level mosaicism detectable be DHPLC but not by direct sequencing. *Hum Mut* 17, 233-234.

14. ***Ponamarev MG, Longley MJ, Nguyen D, Kunkel TA, Copeland WC:*** (2002) Active site mutation in DNA polymerase gamma associated with progressive external optahlmoplegia causes error prone DNA synthesis. *J Biol Chem.* 277, 15225-15228.

15. ***Spelbrink JN, Li FY, Tiranti V, Nikali K, Yuan QP, Tariq M, Wanrooij S, Garrido N, Comi G, Morandi L, Santoro L, Toscano A, Fabrizi GM, Somer H, Croxen R, Beeson D, Poulton J, Suomalainen A, Jacobs HT, Zeviani M, Larsson C:*** (2001) Human mitochondrial DNA deletions associated with mutations in the gene encoding Twinkle, a phage T7 gene 4-like protein localized in mitochondria. *Nat Genet* 28, 223-231.

16. ***Levin BC, Cheng H, Reeder DJ:*** (1999) A Human Mitochondrial DNA Standard Reference Material for Quality Control in Forensic Identification, Medical Diagnosis, and Mutation Detection. *Genomics* 55, 135-146.

17. ***Wallace DC, Brown MD, Lott MT:*** (1999) Mitochondrial DNA variation in human evolution and disease. *Gene* 238, 211-230.

CHAPTER

6

MUTATION ANALYSIS OF LDLR MUTATIONS IN PATIENTS WITH FAMILIAL HYPERCHOLESTEROLAEMIA

Caroline F. Bunn and Peter M. George

Contact:
Caroline F. Bunn
Molecular Pathology, Canterbury Health Laboratories,
PO Box 151, Christchurch 8001, New Zealand
E-mail: cbunn@caverock.net.nz

-9664027-7-4/03/$0.00+$.50 *From:* **Genetic Variance Detection~Nuts & Bolts of DHPLC in Genomics** (pp.117-134)
Edited by: Karl H. Hecker, Ph.D.

6.1 Introduction

In some families, the risk of early coronary heart disease is considerably raised by the inheritance of a specific mutation, in one of two genes (apolipoprotein B, ApoB [OMIM #107730], or low-density lipoprotein receptor, LDLR [OMIM #143890]), which gives rise to Familial Hypercholestrolaemia (FH)[1]. FH is characterized by an autosomal dominant inheritance, severe hypercholesterolaemia, premature onset of atherogenesis and cholesterol deposits in the skin, tendons (tendon xanthoma) or in and around the eyes (corneal arcus and xanthelasma). In the majority of FH patients, the disorder is caused by a mutation in the LDLR gene that destroys or significantly impairs its proper function[2].

FH is among the most common metabolic genetic disorders, the prevalence being estimated at one in 400 to 500 people for mutations in LDLR, and one in 1000 for ApoB mutations[2], more frequent than either cystic fibrosis or sickle cell anaemia. Recent reports from the UK and the Netherlands suggest that the vast majority of these individuals remain unidentified and untreated[3,4]. In both these countries there is extensive evidence that case finding combined with family screening and mutation detection is a very effective method for increasing the percentage of FH individuals getting effective treatment.

With the exception of a few populations where founder mutations have been reported (French Canadians, Finns, Ashkenazi Jews of Lithuanian descent, Christian Lebanese, Dutch and Afrikaners), a relatively large number of diverse mutations cause FH in most populations, with different mutations prevailing in different populations[5-9]. Over 700 LDLR gene mutations have been reported world-wide and new mutations are reported regularly (http://www.ucl.ac.uk/fh/), as opposed to only three mutations that have been described in the apoB gene that cause FH. Since the most frequent known LDLR mutations are present only in a small proportion of patients, an initial strategy of testing for specific mutations (such as is routine for Cystic Fibrosis where a single mutation accounts for more than 70% of mutations) is precluded as a first step in those countries with heterogeneous populations. For some heterozygous FH individuals, a clear diagnosis can be made on the basis of grossly increased cholesterol and associated clinical features. However, one of the problems of using hypercholesterolaemia as a primary criterion for diagnosis of FH is that plasma lipid levels in heterozygous FH patients overlap with those in the general population. The use of DNA tests to confirm the diagnosis of FH in family members of identified LDLR mutation carriers suggests that between 15-20% of adult relatives and 5-10% of children may be misdiagnosed by cholesterol testing alone[4].

The LDL receptor is encoded by a large gene comprised of 18 exons spanning approximately 45 kilobases. Its size, along with the large numbers of mutations found throughout the coding and control regions[10], means that mutations are not 0easily detected by simple diagnostic assays and sequencing the patients' entire LDLR genes on a clinical scale is prohibitive due to cost. By far the majority of mutations described in this gene are single base substitutions or deletions and insertions of only a few base pairs (approximately 90%), with the remainder deletions of multiple introns and exons[2]. In addition, 35 polymorphisms and a number of silent mutations have been described throughout the LDLR gene. In particular, the LDLR polymorphisms complicate the screening process as although the allelic frequencies for these polymorphisms within various populations range from 0.5% to 96%, the majority occurs with allelic frequencies from 30-70 %. Therefore, it is sensible to use a pre-screening method to identify specific small regions of variation for further analysis by sequencing, as has been the case in BRCA1 and BRCA2 gene mutation screening for familial breast cancer[11].

Methods, such as single strand conformational polymorphism (SSCP), previously used to screen for LDLR mutations, lack sensitivity and are labor intensive. The optimal length of DNA for SSCP analysis appears to be from 150 to 200 nucleotides with mutation detection sensitivity for fragments this size ranging from 70-90%[12,13]. The sensitivity of this technique decreases with increasing size of the fragment and it is often necessary to run samples under a number of different experimental conditions, including electrophoresis temperature, in order to detect variants reliably. DHPLC has been reported to have numerous advantages over SSCP; analysis is rapid, inexpensive and semi-automated. Amplicons from 200 to 500 nucleotides are within the ideal range for heteroduplex detection, with sensitivity and specificity reported to range from 96-100 %. In addition, in most cases elution profiles are distinct for any given sequence variant allowing differentiation of polymorphisms from pathogenic mutations[14].

Previously we have reported a comparison of SSCP with DHPLC, assessing their effectiveness as prescreening methods for LDLR mutation detection in New Zealand patients with FH in a research setting[15]. Compared with DHPLC, we were able to detect only 64% of mutations by SSCP. As a result, we have implemented diagnostic testing through LDLR mutation screening by DHPLC. We have developed the first LDLR gene diagnostic assay that integrates DHPLC prescreening with automated PCR setup and DNA sequencing of variants.

6.2 LDLR Mutation Detection in a Diagnostic Setting

A number of factors must be taken into account when implementing a new diagnostic assay to enable timely delivery of an effective service. These include: turn around time for reporting, automation for greater sample numbers and reduced risk of handling errors, optimization to a single set of PCR parameters for reduced handling conditions and ease of automated set-up and, perhaps of prime concern, specificity and sensitivity to avoid reporting of false positives and false negatives.

Analysis of the LDLR gene requires amplification of each patient sample in a number of fragments or amplicons to encompass the coding and control regions. Previously, the LDLR gene has been amplified under multiple PCR conditions in 21 fragments for analysis by SSCP, ranging in size from 127 to 355 base pairs[12,13]. As it was essential that the LDLR diagnostic assay developed met the turnaround time and automation requirements detailed above, the screening process was designed to allow for analysis of patient samples in batch sizes consistent with sample numbers obtained. Therefore, seven patient samples plus controls for all 21 amplicons are screened simultaneously per batch in two microtiter plates (168 wells in total) in a largely automated process.

Samples were obtained from apparently unrelated individuals with clinically probable FH, attending lipid disorders clinics at Christchurch or Dunedin Hospitals. These patients had plasma cholesterol of >8.0 mmol/L and family histories of hypercholestrolaemia and/or classical clinical stigmata of FH. DNA was extracted from patients' whole blood specimens according to an established method[16]. The known mutations in the apoB gene (R3500Q, R3531C and R3500W) were excluded by PCR and restriction analysis. After PCR set-up in two microtiter plates by a Tecan robotic workstation (Tecan AG, Switzerland), all exons were amplified under one set of conditions using Roche Taq Polymerase (Germany). Prior to analysis, heteroduplices were formed by heating the PCR products at 95°C for five minutes and slowly cooling to room temperature over one hour. Samples were analyzed overnight by DHPLC[14] using a WAVE® Nucleic Acid Fragment Analysis System including a C18 reversed phase column based on non-porous poly(styrene-divinylbenzene) particles (DNASep® Cartridge) from Transgenomic (Omaha, NE, USA). DNA was eluted from the column by an acetonitrile gradient in 0.1 mol/L triethylammonium acetate buffer (TEAA), pH 7, at a constant flow rate of 0.9 ml/min. The melting profile for each DNA fragment, the respective elution profiles and column temperatures were determined using the WAVEMAKER™ software from Transgenomic, with further optimization performed against a positive control. Every sequence alteration was confirmed in two independently amplified PCR products by direct cycle sequencing of double-stranded DNA with 33P-labeled terminators and ThermoSequenase according to the manufacturer's instructions (Amersham Pharmacia Biotech, NZ).

Whilst use of a single set of conditions has greatly improved speed of sample preparation for analysis by DHPLC, there have been some disadvantages. Acceptable products are produced but these are not necessarily optimal in terms of yield or quality of product for DHPLC. Some non-specific pre-peaks are observed during DHPLC for certain amplicons, though these are consistent between samples and runs, allowing the operator to correct for their presence. Although PCR yield varies between amplicons, variation is also seen to a similar degree between patient samples (at a factor of between one and three) and can be adjusted for by small changes in scaling of traces.

6.3 Optimization of DHPLC Screening

The LDLR gene is similar to many other genes in terms of approach to mutation screening by DHPLC. Factors for consideration include: primer selection, size of amplicons, choice of positive controls, presence of multiple melting domains within fragments and optimum temperature for analysis. However, the LDLR gene is larger than most genes screened routinely in a diagnostic environment and has a larger number of mutations, with more than 700 described throughout and more being regularly reported. These mutations are also spread throughout all of the exons and intron/exon boundaries, which are both relatively polymorphic. Therefore, there is a distinct chance of DHPLC profiles being similar for different mutations. The influence of these factors on the development of diagnostic LDLR screening is discussed below.

6.3.1 Selection of Primers

A number of primer sets have been described for LDLR mutation analysis by both SSCP and by denaturing gradient gel electrophoresis (DGGE) analysis, which is considered to be more sensitive but also more technically challenging and labor intensive than SSCP (http://www.ucl.ac.uk/fh/primers.html). As we were interested in comparing SSCP and DHPLC, a primer set was selected for both applications based on SSCP design requirements[12]. Whilst DHPLC melting profiles for some of the amplicons obtained by use of these primers were not ideal, they provided an effective starting point for the screening process.

In some cases primers annealed to sequences within or immediately adjacent to intron/exon boundaries and, as approximately 46 splice site mutations have been described for the LDLR[17], it was necessary to relocate some of these primers to encompass these regions. For example, the Exon 2 product, amplified using the primers designed by Hobbs and colleagues, encompasses a stretch of six nucleotides containing 17 reported splice mutations clustered in the 5′ donor site[12]. However, the upstream primer anneals to these nucleotides, rendering detection of any of these mutations unlikely. Therefore a new amplicon was designed to allow analysis of this region, with a new primer annealing approximately 50 nucleotides from the intron/exon boundary.

6.3.2 Generation of Positive Controls

Although various software packages can predict the 'optimal' temperature for screening by DHPLC the use of positive controls, containing a single base substitution or small deletion, is essential to verify and refine the temperature used for each amplicon. These controls also provide some security that the discriminating conditions are maintained between DHPLC runs. Non-identification of abnormal traces for the positive control would indicate variation in buffer conditions, column condition or oven calibration, and is essential in controlling for false negatives during diagnostic screening. Calibration of ovens may also vary between instruments, resulting in variation when screening is performed (at slightly different temperatures) on multiple instruments. This also means that temperatures predicted by the WAVEMAKER™ software, or as reported in the literature, provide a good guide but may not be directly translatable from instrument to instrument. Therefore, temperature optimization and verification by way of positive controls is important and will resolve any issues arising from these differences.

However, obtaining positive controls for each amplicon is a difficult step. Here they were obtained from Professor Steve Humphries (University College London, UK) and Dr. Ros Thiart (University of Stellenbosch), except in the cases of the promoter region and exon 18 (***Table 6-1***). Positive controls were synthesized for the latter two fragments as described below.

A number of methods have been assessed within our laboratory previously for generating positive controls, including: screening of amplicons for polymorphic restriction enzyme sites, site directed mutagenesis (SDM) using the megaprimer protocol and Ligation During Amplification (LDA) SDM[18]. As no polymorphic restriction sites were identified in either the promoter or exon 18 amplicons, it was necessary to generate positive controls for these amplicons. Introduction of a specific mutation through SDM by megaprimer protocol involves a multistep PCR and restriction digest protocol[19], whereas LDA SDM involves a single extension and ligation PCR - allowing more rapid synthesis[20]. Due to rapidity and ease of synthesis, the LDA SDM method was selected as the most suitable method for generation of positive controls and was used to eliminate a unique restriction site within each amplicon.

In brief, a thermostable ligase and a phosphorylated mutagenic primer, containing a single base mismatch, were included in a standard PCR reaction mix and cycling conditions modified to include a seven-minute extension at

65°C. The mutagenic primer was designed to eliminate a unique restriction site, allowing digestion of wild type product post-amplification. Digested reactions were then diluted and re-amplified to obtain 100% mutant product. The mutant product was sequenced to confirm the mutation and was mixed in an equal ratio with wild type PCR product in order to obtain heteroduplices for DHPLC analysis. The synthesized mutant was analyzed by DHPLC against a wild type (normal) control (Figure 6-1). This method allows the generation of multiple positive controls within one to two days.

6.3.3 Multiple Melting Domains

The LDLR gene is relatively GC rich, with several exons having long repetitive stretches of GCs (Table 6-1). Due to sequence constraints, it may not always be possible to design primers to amplify regions that contain similar GC content and therefore have similar melting profiles across the amplicon for DHPLC. In these cases, with referral to the WAVEMAKER™ software predictions for melting domains, it is desirable to screen samples at two temperatures to resolve sequence variation in different domains within an amplicon.

Where possible, multiple positive controls were used in order to optimize partial denaturing temperatures for DHPLC. Two Exon 1 positive controls, localized to different regions of the amplicon, were used during DHPLC optimization (Table 6-1) and further patient screening was performed at both 65°C and 66°C. This example highlights the importance of positive control selection as if only the C>T variant had been used then the optimal screening temperature selected would have been 66°C. Therefore, variants localized in the lower melting domain could be potentially missed if those regions are denatured at 65°C (Figure 6-2).

6.3.4 Similarity of DHPLC Profiles

As discussed previously, a number of factors complicate screening of the LDLR gene for genetic variation that might cause FH. Of considerable importance are the large number of mutations and polymorphisms previously described in the gene and the increasing number of novel mutations being reported, meaning that any LDLR screening process must be able to potentially target any mutation present rather than solely target known variants. As DHPLC has the capacity to detect all these small variations at the DNA level (substitutions, insertions and deletions) without discrimination it is ideally suited to LDLR mutation analysis. However, unlike other genes that might have fewer mutations more sparsely spaced through coding and control regions[13], it is unlikely that elution profiles will be distinct for each LDLR sequence variant making the use of these profiles less effective in discriminating deleterious mutations from benign polymorphisms.

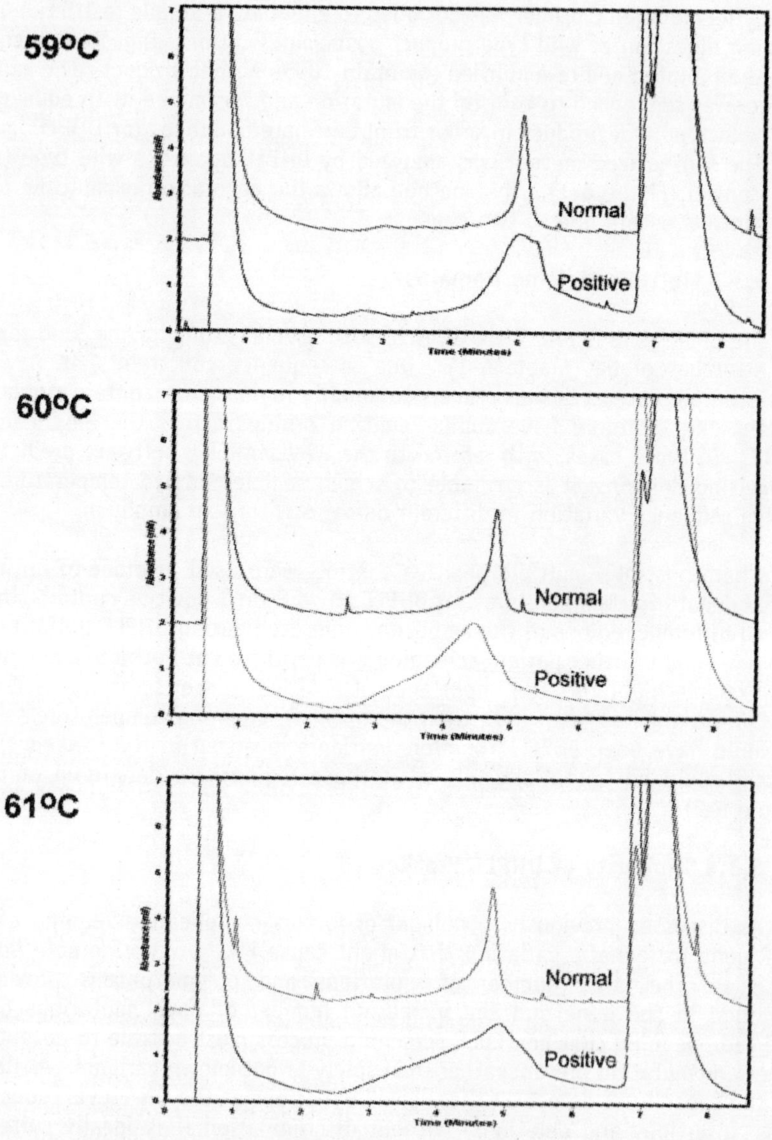

Figure 6-1: Synthesis of a DHPLC positive control for the promoter region of the LDLR gene. Mutant PCR product heteroduplexed with a wild type control was analyzed by DHPLC from 59 to 61°C, with a predicted temperature of 61°C. The mutant trace (Positive) is compared with the wild type control (Normal).

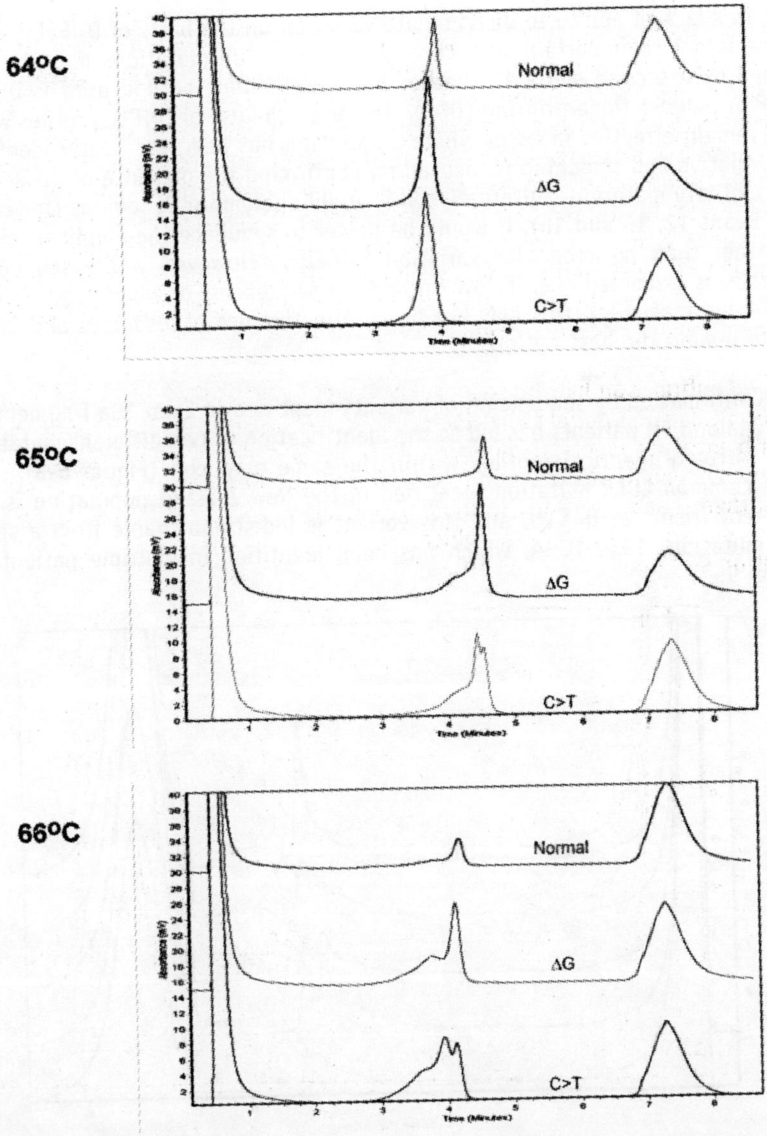

Figure 6-2: Temperature optimization of Exon 1 amplicon. Traces for the two positive controls (ΔG and C>T, traces two and three for each graph, respectively) used to optimize the temperature selection for DHPLC screening are shown from 64 to 66°C. A wild type control (Normal) is included at each temperature.

This lack of ability to discriminate variation on the basis of DHPLC profiles alone is not so important in a clinical context as all mutations must be confirmed by a second method, whether the prescreening method used is DHPLC, SSCP or dideoxy fingerprinting (DDF). However, the use of DHPLC profiles would still remain effective in cases where a mutation has been previously identified in a relative and screening is focused on confirming the presence or absence of that solitary mutation. Alternatively, in highly polymorphic regions (for example, Exons 12, 15 and 18), it would be better to sequence the amplicon directly rather than prescreen for variation initially. However, where sequencing capacity is restricted (i.e. in low throughput sequencers or when sequencing is performed manually), this may not be an effective use of resources and DHPLC would remain the best option.

Confirmation by sequencing of variants identified in Exon 10a fragments of New Zealand FH patients has led to the identification of two different mutations with virtually identical profiles within the same amplicon (Figure 6-3). The most common LDLR mutation identified in the New Zealand population is "FH Northern Irish", or D461N, and this variant is indistinguishable from a splice site mutation, 1359-1G>A, which has been identified in multiple patients in Intron 9.

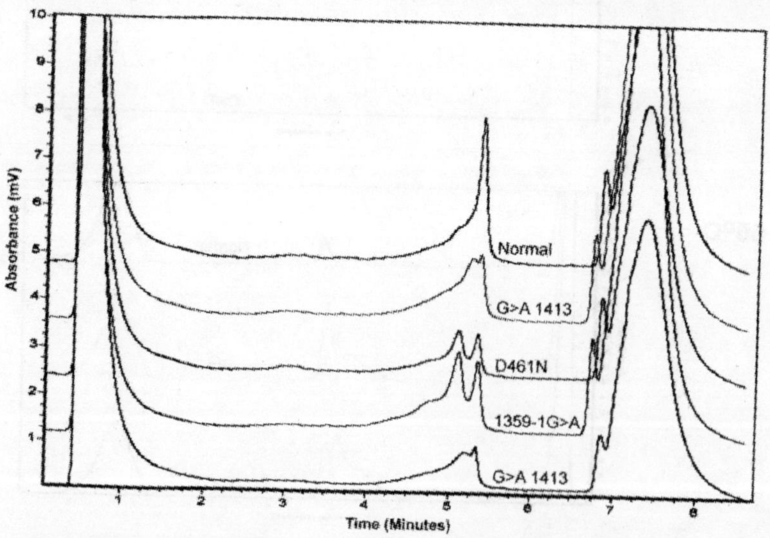

Figure 6-3: DHPLC traces for Exon 10a variants. The wild type control is displayed as the top trace (Normal). Two patients show identical traces for the common polymorphism (G>A 1413; traces two and five). The remaining two very similar traces are representative of the "FH Northern Irish" mutation (D461N: third trace) and a splicing mutation in Intron 9 (1359-1G>A: fourth trace).

6.4 Conclusions

On a worldwide basis it has been estimated that more than 10 million people have FH, of whom as many as 200,000 die of premature coronary heart disease each year[21]. In contrast to most genetic disorders, an effective treatment is available for FH. By using lipid-lowering drugs, a reduction in LDL cholesterol of 50-60% can be achieved, which if started early may be expected to achieve a normal life expectancy. Recent reports from the UK and the Netherlands suggest that the vast majority people with FH remain unidentified and untreated[22,23]. In both of these countries there is extensive evidence that case finding combined with family screening and mutation detection is a very effective method for increasing the percentage of FH individuals getting effective treatment. Therefore an effective means of screening for LDLR mutations, that recognizes that there is a great diversity of LDLR mutations and that targets both known and novel mutations, is needed, especially in heterogeneous populations.

We have shown that DHPLC is an effective tool for the detection of mutations within the LDLR. Twenty-nine different LDLR mutations have been identified in 55 New Zealand FH patients by DHPLC. These mutations are localized throughout the gene, occurring in the receptor's various functional domains (***Table 6-2***). The majority of mutations were identified within the EGF precursor domain (67.4%), followed by the ligand binding domain (23.6%) and the balance are localized in the signal peptide (1.8%), the sugar-modified domain (3.6%), the membrane-spanning domain (1.8%) and the cytoplasmic domain (1.8%).

In the past, screening programs for LDR mutation analysis have focused on "mutation hot spots" such as the ligand-binding domain, specifically Exons 3 and 4 [12,24,25]. However, more recently it has been recognized that this targeting of "hot spots" has resulted in sampling bias rather than being a true reflection of the localization of mutations[15,26-28]. In fact, more novel mutations are being identified within the EGF precursor domain than in any other region, now that more comprehensive screening of the LDLR gene is being performed[26-28]. This is not unexpected, as mutations in this functional domain would affect lipoprotein dissociation from the receptor in endosomes during receptor recycling and receptor positioning at the cell surface[29-31]. This effect of sampling bias highlights the necessity for a comprehensive screening approach that encompasses all coding and control regions, as well as reported and novel mutations.

Mutations were detected in 20.9% of the patients analyzed. The majority of these patients were selected according to total cholesterol levels, >8.0

mmol/L (300mg/dl), and a suggestive family history of heart disease. Less than one third also displayed clinical stigmata. These patients could be categorized into "probable FH" and "definite FH" according to the absence or presence of stigmata, respectively. Mutations have been identified in 45% of patients with "definite FH" compared to 8% of those with "probable FH." These data are consistent with those reported for other populations, which range from 18-27%[32,33], reinforcing that hypercholestrolaemia is not sufficient as the primary criterion for making a clinical diagnosis of FH.

The LDLR gene is an ideal candidate for mutation analysis by DHPLC due to its relatively large size, the range of variation identified throughout the gene and also the nature of FH expression. FH is an autosomal dominantly inherited condition with variable phenotypic presentation. Whilst FH heterozygotes are prone to premature coronary artery disease, homozogous individuals are more severely affected and may die before reaching maturity. However, homozygous affected individuals are rare, having a prevalence of one in one million, compared with one in 400 to 500 for heterozygotes[12,29]. Therefore, FH patients attending lipid clinics will generally be heterozygotes. Thus, unlike those disorders where some samples may be homozygous and must be combined with wild type DNA, mixing of LDLR amplicons is not required for FH.

DHPLC allows a more rapid turnaround time for LDLR mutation screening than SSCP, as amplification of gene fragments and pre-screening by DHPLC is largely automated, less laborious, cheaper and 30% more sensitive than SSCP. Taken together with our finding that it provides sensitive mutation detection of LDLR variants, DHPLC is ideally suited to LDLR mutation analysis in both clinical and research settings.

Table 6-1: Positive controls for optimization of DHPLC

Amplicon	Fragment length (base pairs)	GC Content %	Positive Control (Mutation)
Promoter	191	51	-142G>A (synthesized)
Exon 1	234	63	ØG (also: C>T used initially)
Exon 2	183/196	52/53	C6W
Exon 3	176	56	E80K
Exon 4a	355	63	C139Y
Exon 4b	267	65	D206E
Exon 5	252	55	R232W
Exon 6	174	57	S265R
Exon 7	169	62	R329P
Exon 8	175	61	ØGGGT "FH Nashville"
Exon 9	271	63	V408M
Exon 10a	202	56	D461H
Exon 10b	162	56	P505S
Exon 11	168	54	G528D
Exon 12	209	54	L578S
Exon 13	217	51	M631V
Exon 14	202	64	C660X
Exon 15	246	60	T705I
Exon 16	127	63	S765L
Exon 17	207	53	R793Q
Exon 18	193	54	2574C>A (synthesized)

Table 6-2: Localization of mutations identified by DHPLC. All nucleotide and amino acid numbering refers to that given in the locus specific database (http://www.ucl.ac.uk/fh/). Mutations first reported within the New Zealand population are indicated (*).

Domain	Mutations identified	Number of patients
Signal peptide	G20R*	1
Ligand binding	D26N, D69G, S78X, E80K, W66G, 679-680delAC, 657delC, D245E, E267K, C292Y	13
EGF precursor	G322S, T392M*, R419G*, Y421N*, 1206-1207delCT*, N407K, V408M, 1359-1G>A, D461N, 1478-1479delCT, D504V, 1706-10 G>A, 1943delC*, 1872delC*, C660X	37
Sugar modified	T705I	2
Membrane spanning	P774H*	1
Cytoplasmic	2395-2403del	1

References

1. ***Humphries SE, Galton D, Nicholls P:*** Genetic testing for familial hypercholesterolaemia: practical and ethical issues. *QJM* 1997; 90:169-181.

2. ***Goldstein JL, Hobbs, HH, Brown MS:*** Familial hypercholesterolaemia. In: Scriver CR, Beaudet AL, Sly WS, Valle D (eds): The Metabolic and Molecular Bases of Inherited Disease, ed 7. McGraw, Hill, 1995.

3. ***Bhatnagar D, Morgan J, Siddiq S, Mackness MI, Miller JP and Durrington PN:*** Outcome of case finding among relatives of patients with known heterozygous familial hypercholesterolaemia. *Br Med J* 2000; 321:1497.

4. ***Umans-Eckenhausen MAW, Defesche JC, Sijbrands EJG, Scheerder RLJM, Kastelein JJP:*** Review of first 5 years of screening for familial hypercholesterolaemia. *Netherlands Lancet* 2000; 357:165-168.

5. ***Leitersdorf E, Westhuyzen DRVD, Coetzee GA, Hobbs HH:*** Two common low density lipoprotein receptor gene mutations cause familial hypercholesterolemia in Afrikaners. *J Clin Invest* 1989; 84:954-961.

6. ***Yamamoto T, Davis CG, Brown MS, Schneider WJ, Casey ML, Goldstein JL, Russell DW:*** The human LDL receptor: A cysteine-rich protein with multiple Alu sequences in its mRNA. *Cell* 1984; 39:27-38.

7. ***Ekström U, Abrahamson M, Sveger T, Lombardi P, Nilsson-Ehle P:*** An efficient screening procedure detecting six novel mutations in the LDL receptor gene in Swedish children with hypercholesterolaemia. *Hum Genet* 1995; 96:147-150.

8. ***Orita M, Suzuki Y, Sekiya T, Hayashi K:*** Rapid and sensitive detection of point mutations and DNA polymorphisms using the polymerase chain reaction. *Genomics* 1989; 5:874-879.

9. ***Southern EM:*** Detection of specific sequences among DNA fragments separated by gel electrophoresis. *J Mol Biol* 1975; 98:503-517.

10. ***Heath KE, Gahan M, Whittall RA, Humphries SE:*** Low-density lipoprotein receptor gene (LDLR) world-wide website in familial hypercholesterolaemia: update, new features and mutation analysis. *Atherosclerosis* 2001; 154:243-246.

11. ***Wagner T, Stoppa-Lyonnet D, Fleischmann E, Muhr D, Pages S, Sandberg T, Caux V, Moeslinger R, Langbauer G, Borg A, and Oefner PJ:*** Denaturing high performance liquid chromatography detects reliably BRCA1 and BRCA2 mutations. *Genomics* 1999; 62:369-76.

12. ***Hobbs HH, Brown MS, Goldstein JL:*** Molecular Genetics of the LDL Receptor Gene in Familial Hypercholesterolaemia. *Hum Mutat* 1992; 1:445-466.

13. ***Humphries SE, Gudnason V, Whittall R, Day INM:*** Single-strand conformation polymorphism analysis with high throughput modifications, and its use in mutation detection in familial hypercholestrolaemia. *Clin Chem* 1997; 43:427-435.

14. ***Oefner and Underhill:*** DNA mutation detection using denaturing high-performance liquid chromatography (DHPLC). In: Dracopoli NC, Haines JL, Korf BR, Moir DT, Morton CC, Seidman CE, Seidman JG, Smith DR (eds): Current protocols in human genetics. John Wiley & Sons, 1998.

15. ***Bunn C, Lintott C, Scott R, George P:*** Comparison of SSCP and DHPLC for the detection of LDLR mutations in a New Zealand cohort. *Hum Mutat* 2002 MIB #492 Online.

16. ***Saiki RK, Chang CA, Levenson CH, Warren TC, Boehm CD, Kazazian HH (Jr), Erlich HA:*** Diagnosis of sickle cell anemia and beta-thalassemia with enzymatically amplified DNA and nonradioactive allele-specific oligonucleotide probes. *N Engl J Med* 1988; 319:537-541.

17. ***Peeters AV, Thiart R, de Villiers JN, Jensen HK, Van Gaal LF, Kotze MJ:*** Intronic mutations at splice junctions in the low-density lipoprotein receptor gene. *Mol Cell Probes* 1999; 13:257-260.

18. ***Wang D:*** Detection of polymorphisms in the betaine-homocysteine pathway. *MSc Thesis* 2002.

19. ***Kammann M, Laufs J, Schell J, Gronenborn B:*** Rapid insertional mutagenesis of DNA by polymerase chain reaction (PCR). *Nucleic Acids Res* 1989; 17:5404.

20. ***Sawano A and Miyawaki A:*** Directed evolution of green fluorescent protein by a new versatile PCR strategy for site-directed and semi-random mutagenesis. *Nucleic Acids Res* 2000; 28:E78.

21. ***Gaudet D, Vohl MC, Perron P, Tremblay G, Gagne C, Lesiege D, Bergeron J, Moorjani S, Despres JP:*** Relationships of abdominal obesity and hyperinsulinemia to angiographically assess coronary artery disease in men with known mutations in the LDL receptor gene. *Circulation* 1998; 97:871-877.

22. ***Wittekoek ME, Pimstone SN, Reymer PW, Feuth L, Botma GJ, Defesche JC, Prins M, Hayden MR, Kastelein JJ:*** A common mutation in the lipoprotein lipase gene (N291S) alters the lipoprotein phenotype and risk for cardiovascular disease in patients with familial hypercholesterolemia. *Circulation* 1998; 97:729-735.

23. ***O'Malley JP, Maslen CL, Illingworth DR:*** Angiotensin-coverting enzyme DD genotype and cardiovascular disease in heterozygous familial hypercholesterolemia. *Circulation* 1998; 97:1780-1783.

24. ***Heath KE, Humphries SE, Middleton-Price H, Boxer M:*** A molecular genetic service for diagnosing individuals with familial hypercholesterolaemia (FH) in the United Kingdom. *Eur J Hum Genet* 2001; 9:244-252.

25. ***Reshef A, Nissen H, Triger L, Hensen TS, Eliav O, Schurr D, Safadi R, Gare M, Leitersdorf E:*** Molecular genetics of familial hypercholesterolemia in Israel. *Hum Genet* 1996; 98:581-586.

26. ***Khoo KL, van Acker P, Defesche JC, Tan H, van de Kerkhof L, Heijnen-van Eijk SJ, Kastelein JJ, Deslypere JP:*** Low-density lipoprotein receptor gene mutations in a Southeast Asian population with familial hypercholesterolemia. *Clin Genet* 2000; 58:98-105.

27. ***Gorski B, Kubalska J, Naruszewicz M, Lubinski J:*** LDL-R and Apo-B-100 gene mutations in Polish familial hypercholesterolemias. *Hum Genet* 1998; 102:562-565.

28. ***Garcia-Garcia AB, Real JT, Puig O, Cebolla E, Marin-Garcia P, Martinez Ferrandis JI, Garcia-Sogo M, Civera M, Ascaso JF, Carmena R, Armengod ME, Chaves FJ:*** Molecular genetics of familial hypercholesterolemia in Spain: Ten novel LDLR mutations and population analysis. *Hum Mutat* 2001; 18:458-459.

29. ***Davis CG, Goldstein JL, Sudhof TC, Anderson RG, Russell DW, Brown MS:*** Acid-dependent ligand dissociation and recycling of LDL receptor mediated by growth factor homology region. *Nature* 1987; 326:760-765.

30. ***Jeon H, Meng W, Takagi J, Eck MJ, Springer TA, Blacklow SC:*** Implications for familial hypercholesterolemia from the structure of the LDL receptor YWTD-EGF domain pair. *Nat Struct Biol* 2001; 8:499-504.

31. ***Herz J:*** Deconstructing the LDL receptor-a rhapsody in pieces. *Nat Struct Biol* 2001; 8:476-478.

32. ***Graham CA, McClean E, Ward AJ, Beattie ED, Martin S, O'Kane M, Young IS, Nicholls DP:*** Mutation screening and genotype:phenotype correlation in familial hypercholesterolaemia. *Atherosclerosis* 1999; 147:309-16.

33. ***Goldstein JL and Brown MS:*** Familial Hypercholestrolaemia; in Scriver CR, Beaudet Al, Sly WS, Valle D (eds): The Metabolic Basis of Inherited Diseases. McGraw-Hill, 1989, pp. 1215-1250.

SECTION III

NOVEL AND INNOVATIVE APPLICATIONS OF DHPLC

CHAPTER 7

NOVEL DHPLC BASED STRATEGY FOR IDENTIFICATION AND CHARACTERIZATION OF (BI)-CLONAL T-CELL RECEPTOR TARGETS FOR THE DETECTION OF MINIMAL RESIDUAL DISEASE IN CHILDHOOD ACUTE LYMPHOBLASTIC LEUKEMIA

Udo zur Stadt, Johannes Rischewski, Conny Eckert[1], Katharina Michael, Reinhard Schneppenheim, and Hartmut Kabisch

1) Charite Medical Center, Campus Virchow Klinikum, Children's Hospital, Department of Pediatric Hematology and Oncology

Contact:
Udo zur Stadt

Department of Pediatric Hematology and Oncology, Children's Hospital, University of Hamburg, 20246 Hamburg, Germany
E-mail: zurstadt@uke.uni-hamburg.de

0-9664027-7-4/03/$0.00+$.50 *From:* **Genetic Variance Detection~Nuts & Bolts of DHPLC in Genomics** (pp.137-150)
Edited by: Karl H. Hecker, Ph.D.

7.1 Introduction

Clonal antigen receptor rearrangements are widely used as patient-specific markers for detection of minimal residual disease (MRD) in acute lymphoblastic leukemia (ALL). Since ALL blasts are derived from one single transformed lymphoid precursor cell, junctional regions of immunoglobulin- or T-Cell Receptor (TCR) rearrangements are unique for each leukemia patient. Southern Blot and/or PCR-based techniques allow the identification and characterization of unknown leukemic samples from the time point of diagnosis. Clonal, incomplete TCR- rearrangements are detectable in about 50% of B-cell precursor (BCP) ALL[1]. Several studies demonstrated the applicability of these targets for detection of MRD[2,3].

Southern Blotting identifies only the type of rearrangement without further information about the junctional region. Therefore, only PCR based approaches generally allow for the subsequent analysis of the product by direct sequencing or by sequencing after cloning into a plasmid vector. Direct sequencing is often limited due to impurities from polyclonal background amplification or due to biclonal rearrangements on both alleles of the leukemic sample. Heteroduplex induction followed by acrylamide gel electrophoresis has been described to circumvent these pitfalls[4,5]. Here, PCR products have to be excised and re-amplified for successful downstream analysis. In cases were the resolution of the method used was insufficient, the samples had to be cloned and several clones had to be sequenced.

We recently described the use of partially denaturing high performance liquid chromatography (DHPLC) for identification of clonal TCR- rearrangements[6]. This method permits the analysis of unknown products in a semi-automated way within 10 minutes after the initial PCR without further hands-on work and is highly reproducible using a target specific analysis protocol.

The aim of our recent study was to test feasibility of DHPLC analysis for identification and characterization of incomplete TCR- rearrangements in childhood precursor B-ALL. These rearrangements occur in about 50% of these leukemias and are widely used for MRD analysis. In a substantial number of cases both alleles are rearranged. These biclonal leukemic samples were separated by DHPLC and fractionated for subsequent direct sequencing. The determined sequences were then tested for detection of MRD on the LightCycler system. The method presented here allows us to combine the preparative advantages of DHPLC for characterization of targets with the analytical properties of real time PCR for quantification of MRD.

7.2 Materials and Methods

7.2.1 Patients

Bone marrow (BM) or peripheral blood (PB) from children with ALL at the time of diagnosis was analyzed. Patients were treated according the COALL 92 or 97 protocol[7]. Cell samples were collected after informed consent from the patients or their parents. The diagnosis was based on morphological and immunological examination of BM or PB. Cell samples contained between 40 to 95% malignant cells.

7.2.2 PCR

PCR analysis was performed with standard primers as described previously by Pongers[8]. A germline V 2-D 3-gene rearrangement (without any nucleotide insertions or deletions around the junctional region) results in the amplification of a 150-bp PCR product. For subsequent analyses of initial leukemic DNA samples 50-µl reactions were performed containing 1 U Taq polymerase (Gibco) and 100 ng DNA as well as final concentrations of 2µM of each primer, 3mM MgCl2, and 200 nM of each dNTP. The PCR buffer used was that recommended by supplier of the Taq polymerase. The PCR cycling protocol included a 5 min denaturation step at 95°C, followed by 35 cycles of 30 s at 95°C, 30 s at 55°C, 30 s at 72°C, and was followed by a final extension step of 2 minutes. PCR was performed in 200-µl, ultra-thin PCR tubes (Biozym, Germany) in a T-gradient thermocycler (Biometra, Göttingen, Germany).

7.2.3 Acrylamide Gel Electrophoresis

An aliquot of the final PCR product (1.5µl) was subjected to electrophoresis on an acrylamide gel followed by a rapid silver staining protocol[9]. In brief, a 0.3 mm ultra-thin polyacrylamide was cast onto Gelbond (FMC) films. Samples were run in a discontinuous buffer system. Gels contained 7.1% glycerol and 33 mM Tris-sulfate buffer, pH 9.0, and were run with a buffer strip containing 140 mM Tris-borate, pH 9.0, onto a cooled (8°C) Pharmacia Multiphore System (Amersham Pharmacia, Germany). The whole procedure takes about 1 h and allows the staining of up to 50 samples. Up to 2µl of the 50µl PCR product were used for separation.

7.2.4 DHPLC Analysis

Eight µl of each PCR product were analyzed by DHPLC on a WAVE System from Transgenomic. The shortest and best suitable gradient for analysis of the targets was determined by running universal gradients at different temperatures, starting with 50°C, for several products (Data not shown). An optimized gradient was then used for analysis of the ALL samples (***Table 7-1***).

7.2.6 Fragment Collection and Sequencing

Fragment collection was performed in a two-step procedure. In an initial DHPLC run the time window for the elution of the peaks of interest was determined. During a second injection the samples of interest were collected during the previously defined time interval. Usually, 80 to 100µl of eluent containing the peak of interest was collected. Collection volume is determined by the product of DHPLC buffer flow rate (0.9 ml/min) and collection time. In detail, the fragment collection software was programmed as follows: Using the first run as a reference for specific peak collection, the detection type was set to "whole window modus." The number of collected fractions then depends on the vial volume and corresponds to the time window in the chromatogram over which the fragments are collected. Collected fractions were analyzed in a "result chart" in the WAVEMAKER™ software. A collection table displays each collected vial relative to the corresponding segment of the peak of interest in the elution diagram. Positive fractions were dried under vacuum until complete removal of the eluent. The latter DNA served as template for direct sequencing with the target specific primers used for amplification.

7.2.7 Real Time PCR

Real Time PCR using the LightCycler System (Roche, Germany) was employed to perform quantitative PCR analyses. A germline D 3 TaqMan probe (5′- 6-FAM-ATATCCTCACCCTGGGTCCCATGCCXT-P; X = TAMRA) was used in combination with patient specific ASO primers and a D 3 specific PCR primer (5′-CTGCTTGCTGTGTTTGTCTCCT). Quantification of residual disease was performed in 20 µl reaction volumes using 1 U Platinum Taq Polymerase (Invitrogen) and 2 µl 10x buffer, 200 µM of each dNTP, 100 nM TaqMan probe, 500 nM antisense primer, and 500 nM ASO primer and 5% DMSO. Amplification conditions for quantification were as follows: 8 min initial denaturation followed by 50 cycles with 5 s denaturation and 23 s annealing/extension at the appropriate temperature between 60 and 68°C. Single point fluorescence measurements were performed during the extension step. Serial ten-fold dilutions of initial leukemic cell DNA in DNA from healthy controls was used for the determination of sensitivity limits and for optimization of the RQ-PCR reaction conditions.

7.3 Results

7.3.1 DHPLC Analysis

DHPLC analysis was performed for the most common V 2-D 3 rearrangement. In order to determine the optimal time window for analysis a 20-min acetonitrile (ACN) gradient from 40 to 72% buffer B, with an increase of 2% buffer B per minute was chosen. The positive control elutes under these conditions between 45 and 60% buffer B. Based on these results the gradient conditions described in Table 1 were chosen as general analysis conditions. Higher column oven temperatures do not lead to a better quality of separation. The length of the gradient was selected within a relatively broad time window, which allows the detection of clonal or biclonal samples with different junctional regions. These junctional regions sometimes differ by more than 30 nucleotides (due to deletions and insertions). Characteristic elution profiles are shown in Figure 7-1. For characterization of the specific junctional regions clonal samples were sequenced directly. Biclonal rearrangements elute as a sharp heteroduplex peak one to two minutes before the two homoduplexes. Biclonal samples lead to negative sequencing reactions due to overlapping junctional regions. In order to circumvent cloning or gel excision of these bands we used a novel fragment collection principle based on time-dependent, single-peak collection of DHPLC separated products. Oligoclonal rearrangements were not further analyzed due to their relative instability during therapy and the risk of false negative MRD results[10].

7.3.2 Fragment Collection

Biclonal samples were subjected to a second DHPLC run. After a first run, elution profiles were analyzed on the WAVE system and results were evaluated with the WAVEMAKER software. A typical biclonal fragment reveals 3 peaks. The heteroduplex fraction elutes from the column at a lower acetonitrile concentration than the corresponding homoduplexes. The resulting elution profile serves as a matrix for subsequent fragment collections. The first homoduplex peak of interest was marked with a time window for initialization of the collection software during the next run. In our application, we started fragment collection from the beginning of the first clonal/homoduplex peak to the end of the second clonal/homoduplex peak with an individual fraction volume of 80µl. The collected fractions were dried under vacuum and sequenced directly after reconstitution. Typical result chromatograms and the corresponding sequences from collected fractions are shown in Figure 7-2. ***Table 7-2*** summarizes the results from 10 collected biclonal samples. Sequence variations in the junctional regions are listed and compared to germ line sequences.

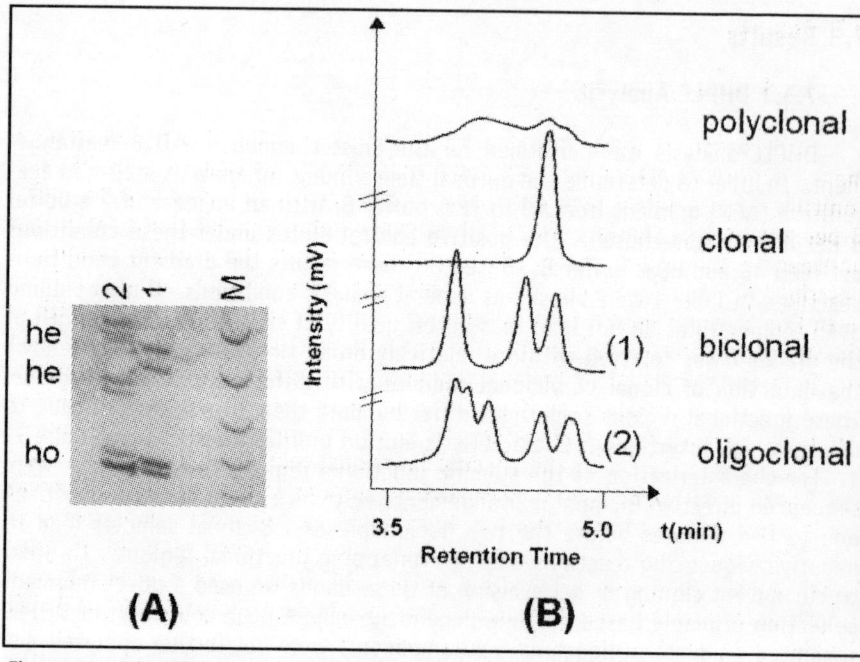

Figure 7-1: Identification of clonal TCR- rearrangements by DHPLC. 8-10µl of PCR product were analyzed on a WAVE System equipped with a DNASep Cartridge at an oven temperature of 50°C. (**a**) Typical acrylamide gel separation of two samples (biclonal and oligoclonal) followed by silver staining, (**b**) characteristic elution pattern are shown for clonal, biclonal, oligoclonal and polyclonal samples.

7.3.3 Real Time PCR

For RQ-PCR a 10-fold dilution series of patient DNA (time point of diagnosis) in DNA derived from 5 healthy unrelated donors was used. The described RQ-PCR approach allowed unequivocal amplification and quantification of both alleles for detection of MRD in each patient (Figure 7-3). Sensitivities of 10-3 to 10-5 were measurable in all samples tested. Thus, both targets from a biclonal rearrangement can be used for a quantitative analysis at different time points during therapy.

7.4 Discussion

The purpose of our study was to evaluate the feasibility of DHPLC for the detection and discrimination of several forms of incomplete TCR- rearrangements (clonal, biclonal and oligoclonal) in childhood precursor B-ALL.

Furthermore, we tested the application of a tandem fragment collector arrangement for purification of biclonal rearrangements for subsequent sequencing reactions. These biclonal PCR products were detectable in about 30% of our V 2-D 3 positive samples (data not shown). All biclonal samples gave a characteristic elution profile and were unequivocally distinguishable from clonal or oligoclonal samples (Figure 7-1), which was in complete agreement with acrylamide gel electrophoresis and heteroduplex analysis results. Whereas clonal rearrangements were sequenced directly, overlapping junctional regions within a biclonal product prohibits this quite simple sequencing approach.

We recently described the feasibility of DHPLC for discrimination of polyclonal from clonal/biclonal TCR- rearrangements[6]. We applied and extended this methodology to the identification of incomplete TCR- rearrangements. V 2-D 3 rearrangements are the most common rearrangements (except IgH) in B-Cell Precursor ALL (BCP-ALL) and therefore widely used for clone specific detection of minimal residual disease[11,12]. Direct sequencing can identify single rearranged fragments[13]. PCR fragments with double rearranged alleles render sequencing inconclusive just at the beginning of the region of interest, i.e.

(a)

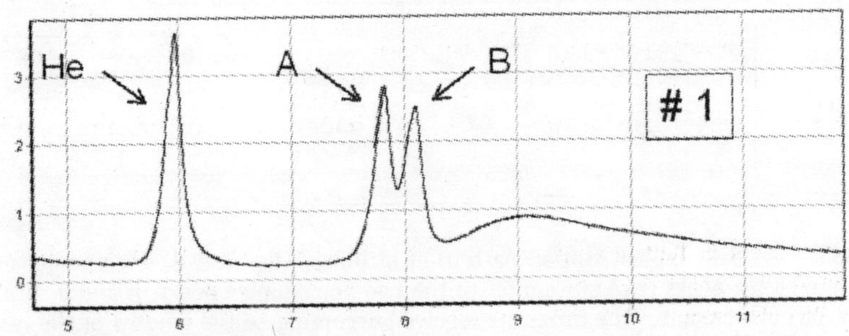

(b)

Vial #	Start [min]	End [min]	Volume [µl]	Sequencing
1	7.65	7.74	79	A
2	7.74	7.83	77	A
3	7.84	7.92	75	-
4	7.92	8.01	75	-
5	8.01	8.10	77	B
6	8.10	8.19	75	B
7	8.19	8.28	75	B
8	8.28	8.35	60	-

(c)

GCACTGTGCGTATCCCCCAGT GGTGTCACAGGCACAGT GL
GCACTGTGCGTATCCCCCAG **GCTAC** TCACAGGCACAGT (A)
-1/+5/-4
CACT **TAAAGGTGACAAGCTCCTTTCT** TGTCACAGGCACAGT (B)
-16/+22/-2

Figure 7-2A

(a)

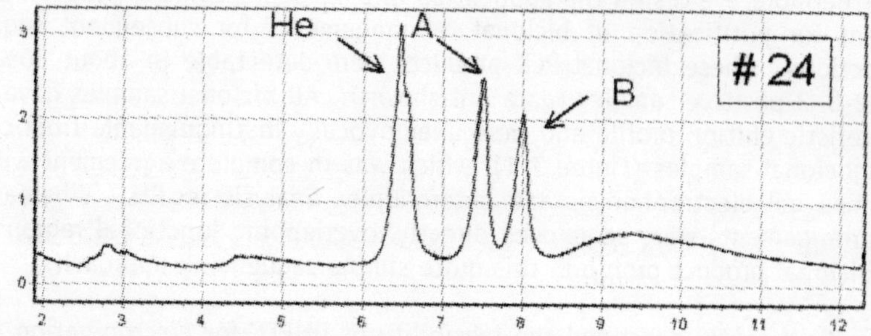

(b)

Vial #	Start [min]	End [min]	Volume [µl]	Sequencing
1	7.32	7.41	80	-
2	7.41	7.50	76	A
3	7.50	7.59	75	A
4	7.59	7.68	77	A
5	7.70	7.76	60	-
6	7.77	7.85	75	-
7	7.86	7.94	77	B
8	7.95	8.03	75	Failure
9	8.03	8.12	75	B
10	8.12	8.19	60	B

(c)

GCACTGTGCGTATCCCCCAGT		GGTGTCACAGGCACAGT	GL
GCACTGTGCGTATCCCCC	**GGGGG** **-3/+5/-8**	AGGCACAGT	(A)
GCACTGTGCGTATCCCCCAG	**GACCC** **-1/+5/-1**	GTGTCACAGGCACAGT	B)

Figure 7-2A

Figure 7-2A & B: Typical elution pattern of biclonal TCR- V 2-D 3 rearrangements. (1) Heteroduplex peaks (He) elute prior to the two homoduplex peaks, A and B, in the case of a biclonal sample. The latter are resolved according to the lengths of the respective PCR fragments. (2) During the second/collection run 8 individual fractions were collected in the time window from 7.65 (7.32) - 8.35 (8.19) min. Up to 80 µl eluent were collected per vial. Five (**a**) and seven (**b**) fractions were sequenced. Sequencing results are assigned to specific peaks observed in the DHPLC chromatogram. (3) Individual sequencing results for peaks A and B are shown relative to the germ line (GL) sequence. Clone specific, junctional regions from each sample were used as allele specific oligonucleotide (ASO) for the detection of MRD.

the junctional region. Compared to the heteroduplex analysis of several other investigators[4,5], DHPLC has an automated setup with no gel preparation or post PCR handling, as we performed no heteroduplex induction in our samples[14]. Additionally, the short and reproducible DHPLC running conditions allow for the analysis of 96 samples over night.

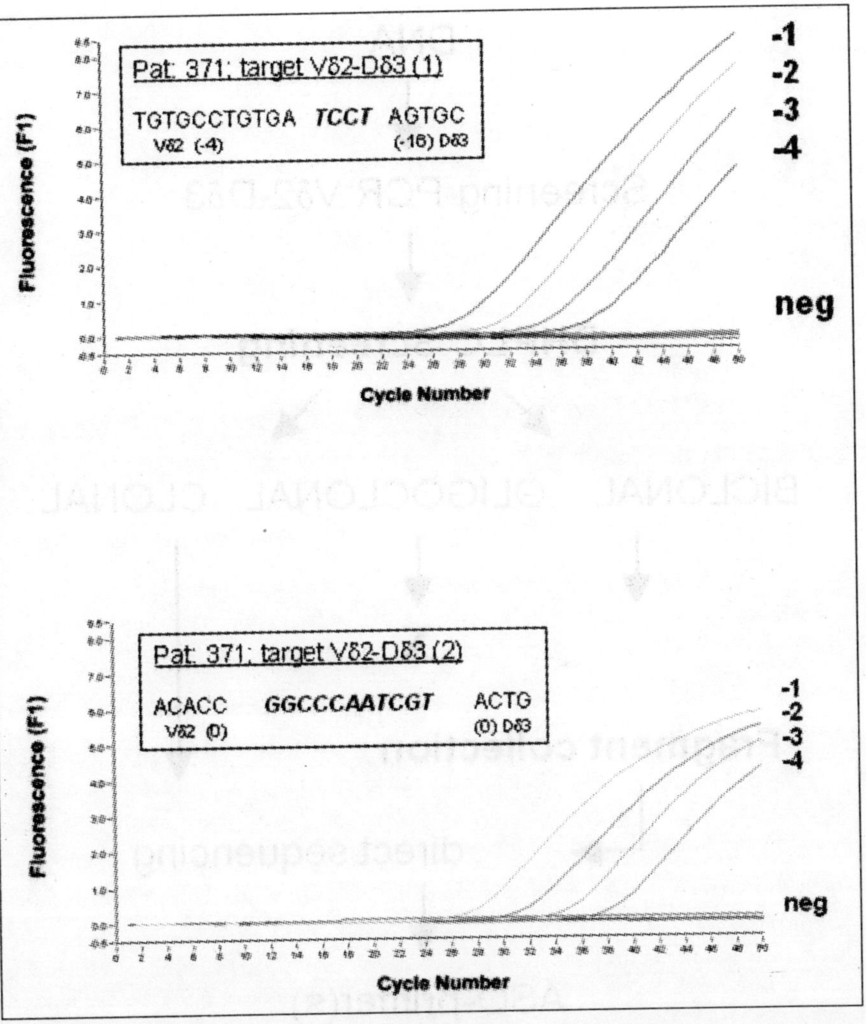

Figure 7-3: RQ-PCR on the LightCycler system. PCR was performed with a germ line Dδ3 PCR primer, a germ line Dδ3 TaqMan probe and a unique ASO patient specific primer. DNA from time point of diagnosis was 10-fold serial diluted in DNA from unrelated samples. 500 ng DNA was used in a 20μl PCR setup. Patient number and corresponding junctional regions contained in the ASO primers (see Table 7-2) are indicated. Diagrams show the fluorescence intensity vs. Cycle number.

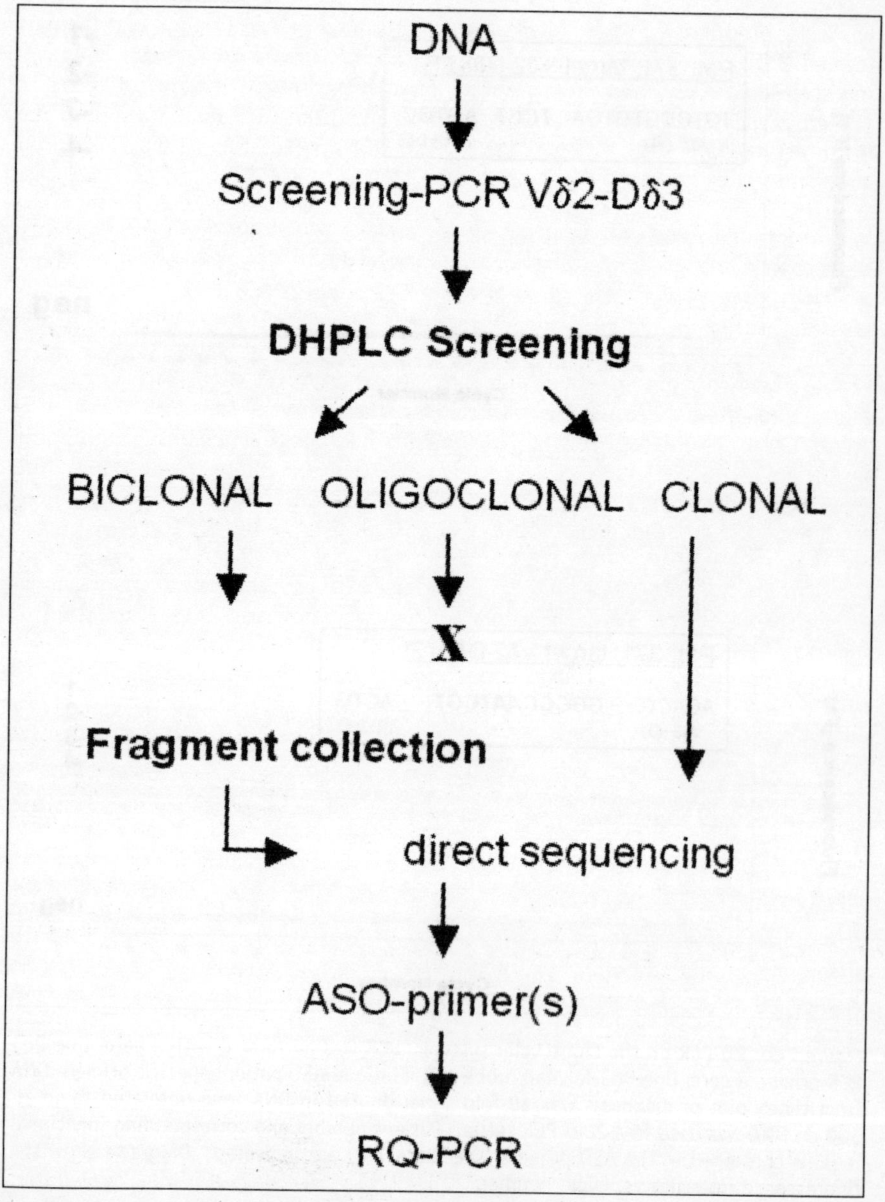

Figure 7-4: Flow chart for the detection, identification, and characterization of TCR-V 2-D 3 targets by means of DHPLC and RQ-PCR.

For a detailed molecular analysis of biclonal products, i.e. sequencing, we performed 2 DHPLC runs for each PCR product. The first run serves as a positioning step for the second collection run. Five to ten fractions were collected on a microtiter plate, 2 to 3 best matching fractions from each peak were prepared for direct sequencing. This procedure allows for a complete analysis of biclonal samples without further PCR or cloning steps.

A detailed molecular analysis of incomplete TCR- PCR products seems to be absolutely necessary for these rearrangements[15,16]. In contrast to Southern blotting analysis, PCR also amplifies minor clonal rearrangements. The resulting small sub-clones are relatively frequent in precursor B-ALL[17]. Steenbergen points out that in precursor B-ALL's this may be a common feature. Sczcepanski et al. clarified this in their comparative study of T- and B-cell receptor rearrangements between initial and relapse diagnosis. Because of the relative instability of oligoclonal targets, incomplete TCR- rearrangements should only be used if they are clonal or biclonal. Especially in oligoclonal samples the response to chemotherapy may vary within different (sub)-clones leading to false negative measurement of minimal residual disease during therapy.

In order to circumvent or minimize these problems we established a novel analytical and preparative procedure for identification and characterization of biclonal rearrangements prior to MRD analysis. RQ-PCR with allele specific oligonucleotides (ASO - primer) corresponding to each allele was performed to test the specificity of the determined sequences. This allows one to use both allelic variants for the exact examination of minimal residual disease during therapy. Figure 7-4 describes a possible procedure for the identification, characterization and application of V 2 3 targets for MRD analysis.

7.5 Conclusion

We present a novel DHPLC-based method for identification of (bi)clonal V 2-D 3 rearrangements in childhood ALL. Biclonal rearrangements with length differences in the junctional region down to 4 base pairs, compared to germ line sequences, were resolved on the DHPLC column. This high peak resolution allows us to purify biclonal samples for direct sequencing using an automated fragment collector.

Our results show: (1) DHPLC analysis for clonality assessment is a rapid and reliable tool for identification of MRD targets in newly diagnosed leukemias under standardized conditions. (2) Fragment collection simplifies the identification of single alleles within biclonal rearrangements. (3) Subsequent real time PCR on the LightCycler now enables us to measure MRD quantitatively. The described strategy of DHPLC followed by RQ-PCR will standardize the molecular detection of MRD in childhood ALL.

Table 7-1: WAVE System running conditions for the separation of incomplete TCR-δ rearrangements.

Step	Time	% Buffer A	% Buffer B	% Buffer C
Loading	0.0	70	30	0
Start Gradient	0.5	55	45	0
Stop Gradient	10.5	40	60	0
Start Clean	10.6	0	0	100
Stop Clean	11.1	0	0	100
Start Equilibrate	11.2	70	30	0
Stop Equilibrate	12.1	70	30	0

Buffer A: 0.1 M TEAA; *Buffer B:* 0.1 M TEAA/25 % ACN; *Buffer C* (wash buffer): 75 % ACN. Runs were performed with a flow rate of 0.9 ml per minute and 1.5% per minute increases in buffer B within the analysis portion of the gradient.

Table 7-2: Examples of direct sequencing results for biclonal samples after DHPLC analysis and fragment collection.

DNA-#	Dδ3 Deletions	„N" - insertion	Vδ2 Deletions	Length to germ line	RQ - PCR Sensitivity
1	-1	GCTAC	-4	0	
	-16	TAAAGGTGACAAGCTCCTTTCT	-2	+4	
24	-3	GGGGG	-8	-6	
	-1	GACCC	-5	-1	
218	-3	GAG	-13	-13	
	-3	TCGGTGTCA	-6	0	
220	-23	ACAGGTCGACCCCCCGC	-3	-9	
	-1	CCTGGAGTC	-7	1	
241	-8	TCCCCTGACAC	-5	-2	-4
	0	TTCCGCCCGGGGG	-5	-8	-5
273	-4	TAAGGAGGC	-8	-3	
	-2	CCG	0	1	
371	-16	AGGA	-4	-16	-4
	0	ACGATTGGGCC	0	11	-4
430	-4	AAGGT	-14	-13	
	0	GAGA	-3	-1	
481	-5	TCTCGGG	-12	-10	-3
	-1	ATA	-3	-1	-4
526	-5	TCGCGGA	-3	-1	-4
	0	CCC	0	+3	-4

Acknowledgement

This work was supported by the "Fördergemeinschaft Kinderkrebszentrum Hamburg"

References

1. ***Breit TM, Wolfers-Tettero ILM, Beishuizen A, Verhoefen M-AJ, van Wering E, van Dongen JJ:*** Southern Blot Patterns, Frequencies, and Junctional Diversity of T-Cell Receptor-delta Gene Rearrangements in Acute Lymphoblastic Leukemia. *Blood* 1993, 82:3063-3074.

2. ***Cave H, van der Werff ten Bosch J, Suciu S, Guidal C, Waterkeyn C, Otten J, Bakkus M, Thielemans K, Grandchamp B, Vilmer E:*** Clinical significance of minimal residual disease in childhood acute lymphoblastic leukemia. European Organization for Research and Treatment of Cancer—Childhood Leukemia Cooperative Group [see comments]. *N Engl J Med* 1998, 339:591-598.

3. ***van Dongen JJ, Seriu T, Panzer-Grumayer ER, Biondi A, Pongers-Willemse MJ, Corral L, Stolz F, Schrappe M, Masera G, Kamps WA, et al.:*** Prognostic value of minimal residual disease in acute lymphoblastic leukaemia in childhood [see comments]. *Lancet* 1998, 352:1731-1738.

4. ***Langerak AW, Szczepanski T, van der Burg M, Wolvers-Tettero IL, van Dongen JJ:*** Heteroduplex PCR analysis of rearranged T cell receptor genes for clonality assessment in suspect T cell proliferations. *Leukemia* 1997, 11:2192-2199.

5. ***Bottaro M, Berti E, Biondi A, Migone N, Crosti L:*** Heteroduplex analysis of T-cell receptor gamma gene rearrangements for diagnosis and monitoring of cutaneous T-cell lymphomas. *Blood* 1994, 83:3271-3278.

6. ***zur Stadt U, Rischewski J, Schneppenheim R, Kabisch H:*** Denaturing HPLC for Idenfication of Clonal T-Cell Receptor gamma Rearrangements in Newly Diagnosed Acute Lymphoblastic Leukemia. *Clinical Chemistry* 2001, 47:3003-2011.

7. ***Janka-Schaub G, Harms DO, Den Boer ML, Veermann AJP, Pieters R:*** In Vitro Drug Resistance as Independent Prognostic Factor in Study COALL-05-92 for Treatment of Childhood Acute Lymphoblastic Leukemia: Basis for Treatment Stratification According to Traditional Risk Factors and Drug Sensitivity in Study COALL-6-97. *Klinische Pädiatrie* 1999, 211:233-238.

8. ***Pongers-Willemse MJ, Seriu T, Stolz F, d'Aniello E, Gameiro P, Pisa P, Gonzalez M, Bartram CR, Panzer-Grumayer ER, Biondi A, et al.:*** Primers and protocols for standardized detection of minimal residual disease in acute lymphoblastic leukemia using immunoglobulin and T cell receptor gene rearrangements and TAL1 deletions as PCR targets: report of the BIOMED-1 CONCERTED ACTION: investigation of minimal residual disease in acute leukemia. *Leukemia* 1999, 13:110-118.

9. ***Budowle B, Chakraborty R, Guisti AM, Eisenberg AJ, Allen RC:*** Analysis of the VNTR Locus DIS80 by the PCR Followed by High Resolution PAGE. *Am J Hum Genet* 1991, 48:137-144.

10. ***Szczepanski T, Willemse MJ, Brinkhoff B, van Wering ER, van der Burg M, van Dongen JJ:*** Comparative analysis of Ig and TCR gene rearrangements at diagnosis and at relapse of childhood precursor B-ALL provides improved strategies for selection of stable PCR targets for monitoring of minimal residual disease. *Blood* 2002, 99:2315 - 2323.

11. ***Seeger K, Taube T, Eckert C, Hanel C, Pogoda M, Henze G:*** Unusual TCR-delta rearrangement patterns revealed by screening of a large series of childhood acute lymphoblastic leukemia by multiplex PCR. *Br J Haematol* 2001, 2001:318-322.

12. ***Szczepanski T, Beishuizen A, Pongers-Willemse MJ, Hahlen K, Van Wering ER, Wijkhuijs AJ, Tibbe GJ, De Bruijn MA, Van Dongen JJ:*** Cross-lineage T cell receptor gene rearrangements occur in more than ninety percent of childhood precursor-B acute lymphoblastic leukemias: alternative PCR targets for detection of minimal residual disease. *Leukemia* 1999, 13:196-205.

13. ***Langlands K, Eden B, Micallef-Eynaud P, Parker AC, Anthony RS:*** Direct sequence analysis of TCR Vdelta2-Ddelta3 rearrangements in common acute lymphoblastic leukemia and application to detection of minimal residual disease. *Br J Hematol* 1993, 84:648-655.

14. ***Xiao W, Oefner PJ:*** Denaturing high-performance liquid chromatography: A review. *Hum Mutat* 2001, 17:439-474.

15. ***Steward CG, Goulden NJ, Katz F, Baines D, Martin PG, Langlands K, Potter MN, Chessells JM, Oakhill A:*** A polymerase chain reaction study of the stability of Ig heavy-chain and T-cell receptor delta gene rearrangements between presentation and relapse of childhood B-lineage acute lymphoblastic leukemia. *Blood* 1994, 83:1355-1362.

16. ***Lo Nigro L, Cazzaniga G, Di Cataldo A, Pannunzio A, D'Aniello E, Masera G, Schiliro G, Biondi A:*** Clonal stability in children with acute lymphoblastic leukemia (ALL) who relapsed five or more years after diagnosis. *Leukemia* 1999, 13:190-195.

17. ***Steenbergen EJ, Verhagen OJ, van Leeuwen EF, van den Berg H, von dem Borne AE, van der Schoot CE:*** Frequent ongoing T-cell receptor rearrangements in childhood B- precursor acute lymphoblastic leukemia: implications for monitoring minimal residual disease. *Blood* 1995, 86:692-702.

CHAPTER 8

SIRPH: A RAPID, QUANTITATIVE SNuPE-IP RP HPLC METHOD FOR SIMULTANEOUS METHYLATION ANALYSIS AT MULTIPLE CpG SITES

Osman El-Maarri[1], Johannes Oldenburg[1,2] and Jörn Walter[3]

1. Institute of Experimental Haematology and Transfusion Medicine, Bonn, Germany;
2. Institute of Transfusion Medicine and Immune Haematology, Frankfurt, Germany;
3. Universität des Saarlandes, Genetik, Saarbrücken, Germany

Contact:
Osman El-Maarri

Institute of Experimental Haematology and Transfusion Medicine,
Sigmund-Freud Straße 25, 53105 Bonn, Germany
E-mail: osman.elmaarri@ukb.uni-bonn.de

0-9664027-7-4/03/$0.00+$.50 *From:* **Genetic Variance Detection~Nuts & Bolts of DHPLC in Genomics** (pp.151-162)

Edited by: Karl H. Hecker, Ph.D.

8.1 Introduction

8.1.1 Bisulfite Based DNA Methylation Analysis

The most widely used protocols for detailed DNA methylation analyses are based on the use of bisulfite converted genomic DNA. This methodology was introduced by M. Frommer in the early 1990's[1] providing the first highly sensitive and sequence independent approach to detect CpG (mammals) and CpG and CpNpG (plants) methylation, independent of the sequence context. The method uses the ability of sodium bisulfite to selectively interact with cytosines at their carbon-6 position to form sulphonated cytosine intermediates. These intermediates are then converted to uracil by pH-dependent deamination and desulphonation steps. 5-methyl cytosine remains un-reactive under such conditions since the presence of the methyl group inhibits the formation of sulphonation at the carbon-6 position[1,2]. The uracil in the bisulfite converted DNA is replaced by thymine in the subsequent PCR reaction. After cloning and sequencing of such bisulfite generated PCR products methylated cytosine will read as cytosine while un-methylated cytosine will read as thymine. Several modifications and improvements of the original protocol have been introduced[3]. A detailed protocol allowing the use of smallest amounts of starting material was published recently[4].

8.1.2 Overview of methods for the analysis of PCR products derived from bisulfite-treated genomic DNA

Rapid and reliable quantification of methylated cytosines in a given genomic region is the main goal in DNA methylation analysis. The traditional method of cloning and sequencing of PCR products derived from bisulfite-treated genomic DNA provides the most detailed information. This approach, however, is very time consuming, laborious and expensive. For many applications the analysis of methylation at selected CpG sites will be sufficient. Since there is a great demand for inexpensive, rapid, and quantitative methods for performing high throughput screening of DNA methylation for diagnostic purposes, several alternative methods have been developed in recent years. One such method is COBRA, which is based on restriction enzyme digestion of PCR products derived from bisulfite treated genomic DNA[5]. The use of this method, however, is limited since it allows only the analysis of CpG methylation within (newly generated) restriction sites of the PCR products derived from bisulfite-treated genomic DNA. Recently, a more flexible method based on differential hybridisation of PCR fragments from bisulfite treated genomic DNA, using oligonucleotide-containing chips, was introduced[6]. While this method allows for a high throughput screening, it requires high technological laboratory standards and sophisticated and laborious chip-design and analysis tools. More flexible methods are based on single nucleotide primer extension techniques (SNuPE techniques) such as: (1) MethyLight, a real time PCR based SNuPE method, which is quantitative and highly sensitive, but which requires special fluorescently labelled primers[7]; (2) Ms-SNuPE which uses the incorporation of radioactive nucleotides that are quantified by autoradiography, requiring separation of extension products on acrylamide gels[8]. Recently, we have developed an inexpensive, non-radioactive variation of such a SNuPE protocol using ion pair reversed phase high performance liquid chromatography (SIRPH: SNuPE-IP RP HPLC) for methylation detection and quantification[9].

8.1.3 SIRPH - Single Nucleotide Primer Extension with Ion Pair Reversed Phase High Performance Liquid Chromatography

PCR products, derived from bisulfite treated DNA, of a region of interest are obtained using standard protocols. We recommend using agarose imbedded DNA material to obtain optimal results[4]. The PCR products are then purified (see protocol below) to remove residual PCR primers and dNTPs. Subsequently primer(s) immediately 5′ to a CpG site(s) are hybridised to the denatured single-stranded PCR product. Temperature cycling using Thermo Sequenase™ in the presence of both ddCTP and ddTTP extends the annealed primers; ddTTP (unmethylated CpG) or ddCTP (methylated CpG, see Figure 8-1) extended products are subsequently analysed by DHPLC directly (WAVE® System, Transgenomic). Due to the incorporation of the more hydrophobic ddTTP the retention time of such an extended product is longer compared to products containing ddCTP (Figure 8-1). The amount of the ddTTP and ddCTP extended products can then be quantified by either integrating peak areas or measuring peak heights and calculation of percentage ratios using WAVEMAKER™ Software version 4.1 (Transgenomic, Omaha, NE) - see analysis section below.

8.2 Materials and Methods

8.2.1 Purification of PCR Products Derived from Bisulfite-treated Genomic DNA

PCR products can be purified either by gel extraction, e.g.: QIAquick Gel Extraction Kit (Qiagen, Valencia, CA), or by treatment with a mixture of exonuclease I and shrimp alkaline phosphatase, e.g.: ExoSap-IT (Amersham), which will degrade excess primer and unincorporated dNTPs.

8.2.2 SNuPE Reaction

Thermo Sequenase DNA Polymerase, 2′,3′-dideoxythymidine 5′-triphosphate (ddTTP), and 2′,3′-dideoxycytidine 5′-triphosphate (ddCTP) were purchased from Amersham Biosciences.

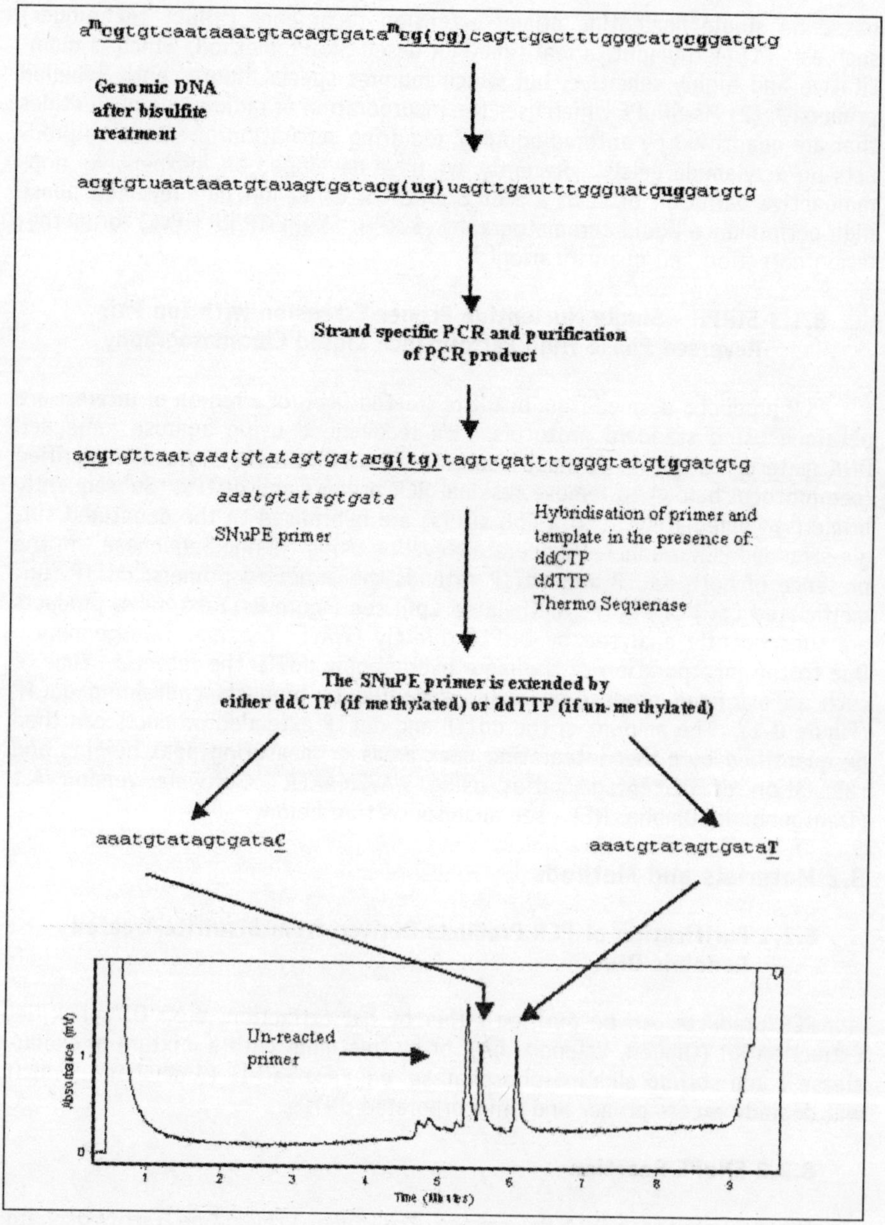

Figure 8-1: General outline of the SNuPE reaction scheme.

8.2.3 IP RP HPLC

All DHPLC analyses were performed on a WAVE® Nucleic Acid Fragment Analysis System (Transgenomic) equipped with a DNASep® Cartridge (Transgenomic). The standard dual buffer system was used with buffer A being 0.1 M aqueous triethylammonium acetate (TEAA) and buffer B being 0.1 M aqueous TEAA with 25% (v/v) acetonitrile. Acetonitrile was obtained from ROTH and HPLC grade water was purchased from Merck.

8.2.4 Protocol Outline

The SIRPH protocol for quantitative methylation analysis at specific CpG sites can be divided into three steps. The first step involves the generation of PCR products derived from bisulfite-treated genomic DNA. The second step comprises the SNuPE reaction, while the third step involves the separation and quantification of primer extension products with the WAVE System. Recently, we published a detailed description of the first step[4]. In the following we describe in detail the latter two steps.

8.2.5 Detailed Protocol for the Analysis of Methylated DNA

Treatment of isolated genomic DNA or cells and PCR is performed as described in the protocol by Hajkova *et al*[4]. The PCR product is purified by either of two methods. (1) Run the PCR product on 1% agarose gel until optimum separation, excise the specific band of interest and recover the product by using a standard PCR-gel extraction kit. In our experience the QIAquick Gel Extraction Kit from Qiagen yields highly pure products with high rate of recovery. (2) Add 2µl of exonuclease I (10 units/µl) and shrimp alkaline phosphatase (2 units/µl) (ExoSap-IT from Amersham) to 5µl of PCR product, heat at 37°C for 15 min followed by 15 min at 80°C to deactivate the enzyme mixture.

Enzymatic treatment has the advantage that it is rapid and easier to perform, however, it is relatively more expensive. The gel extraction on the other hand is more laborious, but it has the advantage of concentrating low abundance PCR products. In cases were non-specific or spurious PCR products are present, these may be removed by this method. However, we found that the enzymatic treatment is very convenient and easy to perform, especially when analysing large numbers of samples.

SNuPE-reaction sample setup (20µl total reaction volume):

Reaction buffer:	2µl	(10 X buffer)
SNuPE primer(s) (***n***):	1µl	(12.5 pmol for each primer)
Template DNA:	1-5µl	(50 to 100 ng)
ddCTP:	1µl	(1 mM solution)
ddTTP:	1µl	(1 mM solution)
Thermo Sequenase:	1µl	(1 unit/µl stock concentration)
H_2O:		bring to final volume of 20 µl

n = the number of primers used for multiplex per reaction

The SNuPE PCR program is performed as follows. Step 1: 94°C for 2 min; Step 2: 92°C for 10 s; Step 3: ***X***°C for 30 s; Step 4: 60°C for 1 min; repeat steps 2 to 4 50 times.

X = the optimum annealing temperature for the primer(s) used (see below: Analysis of results).

8.2.6 HPLC Analysis

Aliquots (10µl) of the above primer extension reactions are loaded directly onto the WAVE® Nucleic Acid Fragment Analysis System (Transgenomic). The oven temperature is set to 50°C. The elution gradient used is shown in ***Table 8-1*** (also see Figure 8-2). The values of B1 and B2 were defined empirically for each set of SNuPE primers. As a first step, the set of primers used is run with a wide gradient of 10% (B1) to 60% (B2) buffer B. Most short primers of 10 to 20 bp should be eluted using this gradient. At a later stage, depending on the retention times of the primers, the gradient can be narrowed down/shortened by either increasing the starting percentage of buffer B, lowering the final percentage of buffer B, or both. Resolution can be adjusted by changing the slope of the gradient. A decrease in slope, the reduction of the change of buffer composition in percent over a fixed time interval, generally translates into better resolution, while increasing gradient slope generally leads to decreased resolution.

8.2.7 Design Guidelines for SNuPE Primers

The most important step of the SIRPH method is the selection and design of suitable sets of primers. Some recommendations for designing appropriate primers are listed below:

(a) The 3′-end of the SNuPE primer has to be just 5′ (flanking) to the specific CpG site to be studied.

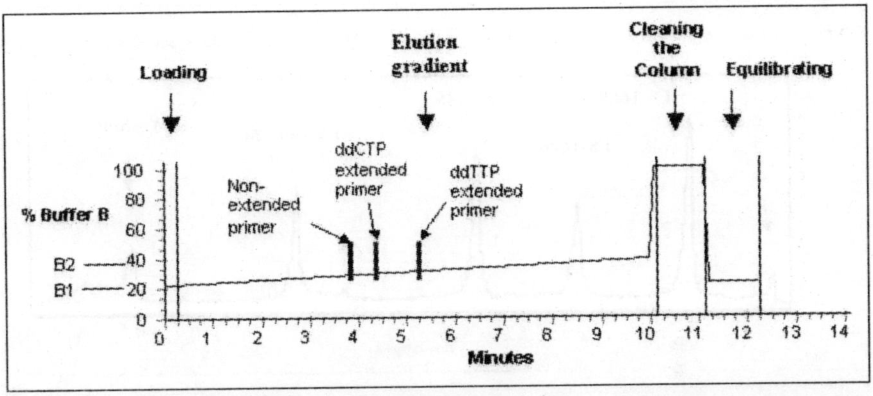

Figure 8-2: Schematic diagram of a 4-step gradient profile. B1 and B2 are starting and ending percentages of buffer B in the elution gradient. These are empirically determined for each set of primers. A theoretical example is also shown were the un-extended excess starting primer, black bar at ca. 3.9 min, is expected to be eluted first at a lower % B than the corresponding extended primers, followed by the ddCTP extended primer (second bar) and finally by ddTTP extended primer (bar at ca. 5.3 min).

(b) Avoid placing primers (if possible) in T-rich regions since this could increase mispriming and lead to inaccuracy in the methylation measurements.

(c) Primers must not include a CpG site, as this will bias measurements.

(d) Avoid designing excessively short primers, as these would have a higher chance for mispriming. In our experience primers as short as 10 bases can give accurate measurements. However, for routine uses our primers are 15 to 18 bases long.

(e) Primers that will be used in multiplex SNuPE reactions should be matched for their melting temperature. However, primers that differ by up to 10 °C in their optimum annealing temperature can be used without apparent decrease in the accuracy of measurements.

(f) For multiplexed SNuPE reactions that are to be analyzed by DHPLC, the retention times of the individual primers and their elongation products must exhibit differences. A general guideline for a rough estimation of relative retention times for two primers is given in ***Table 8-2*** (see also Figure 8-3). If for practical reason two primers have to be designed that give similar retention times, we recommend to extend one of the primers by the addition of thymidins (T) to its 5´-end. In our experience, this addition has no functional effect on

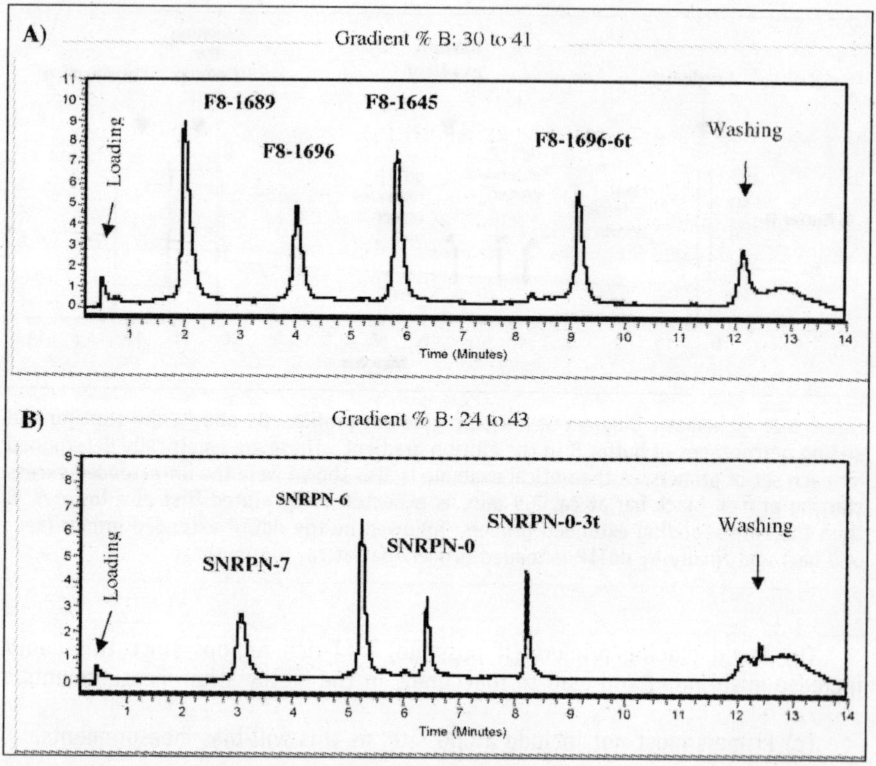

Figure 8-3: Chromatograms of multiplexed SNuPE primers showing the relative retention times as a function of the primers' base composition. A) F8 primers and extension products. B) SNRPN primers and extension products. For an estimation of the elution order of primers and their extension products see Table 8-2.

either the annealing temperature to the template, or on the SNuPE reaction itself. The number of T to be added has to be determined empirically. Each additional T leads to an incremental, stepwise increase of the retention time observed.

8.3 Analysis of results

The percentage of the methylated portion of genomic DNA before bisulfite treatment can be calculated from the chromatograms of the SNuPE products according to the formula: M = (AC/(AC+AT)) * 100, were AC and AT are the peak areas or peak heights of ddCTP and ddTTP extended primers, respectively. Use of peak heights yields more precise results than integrated peak areas. The latter, however, still yields a very good estimate of the percentage of methylation present. WAVEMAKER™ software automatically calculates the AC and AT peak areas and heights.

8.4 Advantages and Drawbacks of SIRPH Analyses

The methylation analysis presented in this chapter has the advantages of being quantitative, rapid, non-radioactive, and suitable for automation. Moreover, a complete picture of the methylation states at a given CpG site in the PCR product under investigation is obtained. However, some drawbacks of the method have to be considered:

(a) Detailed sequence information of the region to be studied is required

(b) In contrast to the cloning and sequencing approach, methylation analysis of only specific CpG sites can be performed with each primer.

(c) Mosaic patterns of methylation, were at given CpG sites random patterns of methylation occur on the same chromosome, cannot be detected.

(d) Imprinted patterns of methylation, with about half of the PCR product being representative of completely methylated genomic DNA and the other half of completely un-methylated DNA, cannot be revealed directly, but have to be deduced from the level of methylation at only few sites.

(e) In a CpG island where CpG sites are very close to each others, i.e.: between 0 and 10 bases, some CpG sites cannot be analyzed because of length restrictions on the primers used in SNuPE reactions.

8.5 Conclusion

We described a detailed protocol for the SIRPH method for quantitative methylation analysis at specific CpG sites. The WAVE® Nucleic Acid Fragment Analysis System (Transgenomic) proved to be adequate for this analysis. The SIRPH assay can be optimized to analyze more than one CpG sites simultaneously and is highly reproducible. The assay is highly beneficial in comparing large numbers of samples in a relatively short time.

Acknowledgments

Part of this study was supported by a BONFOR grant (O-145.0004). We thank Joachim Fisher and Sven Langner from Transgenomic for valuable discussions and technical support and Ursula Herbiniaux for technical support.

Table 8-1: WAVE System elution gradient for SNuPE-product analysis.

Step	Time	% Buffer A	% Buffer B
Loading	0.0	100 - B1	B1
Start Gradient	0.1	100 - B1	B1
Stop Gradient	10.0	100 - B2	B2
Start Clean	10.1	0	100
Stop Clean	11.1	0	100
Start Equilibrate	11.2	100 - B1	B1
Stop Equilibrate	12.2	100 - B1	B1

Buffer A is an aqueous solution of 0.1 M TEAA. Buffer B is an aqueous solution of 0.1 M TEAA containing 25 % (v/v) acetonitrile (ACN). B1 and B2 are starting and ending percentages of buffer B in the elution gradient.

Table 8-2: Two examples of how to roughly estimate the relative retention time of primers. Example A is the from factor VIII exon 14 region, while example B is from the SNRPN exon 1 region.

	Primer Sequence	Actual retention time[a]	Nucleotide contents[b]			Estimated Retention Index[c]
			G x1	A x2	T x2.5	
Exemple A	F8-1689: gat gaa aat tag agt ttt	1.98	4	14	17.5	35.5
	F8-1696: agt ttt taa aag aaa ata	4.02	2	20	15	37
	F8-1645: aat tta tta gtt ttg aaa	5.82	2	14	22.5	38.5
	F8-1696-6t:ttt ttt agt ttt taa aag aaa ata	9.12	2	20	30	52
Exemple B	SNRPN-7: ga tag ttt ggg gag	3.08	7	6	10	23
	SNRPN-6: taa ggt tag ttg tgt	5.24	5	6	17.5	28.5
	SNRPN-0: ggg att ttt gta ttg	6.42	5	4	20	29
	SNRPN-0-3t: ttt ggg att ttt gta ttg	8.20	5	4	27.5	36.5

a These are the actual experimental values shown in the chromatogram in Figure 8-2.
b The nucleotide content is defined as the number of G's times 1, A's times 2, and T's time 2.5. The multiplication factors are largely empirically deduced and they are reflections of the relative hydrophobicity of the individual nucleotides.
c The estimated retention factor (ERI) is calculated by adding the nucleotide content for Gs, As and Ts over the length of the primer. This is particularly useful to predict the order of the elution of primers that have the same number of bases.

References

1. ***Frommer M, McDonald LE, Millar DS, Collis CM, Watt F, Grigg GW, Molloy PL, Paul CL:*** A genomic sequencing protocol that yields a positive display of 5-methylcytosine residues in individual DNA strands. *Proc Natl Acad Sci U S A* 1992; 89:1827-31.

2. ***Raizis AM, Schmitt F, Jost JP:*** A bisulfite method of 5-methylcytosine mapping that minimizes template degradation. *Anal Biochem.* 1995; 226:161-6.

3. ***Olek A, Oswald J, Walter J:*** A modified and improved method for bisulphite based cytosine methylation analysis. *Nucleic Acids Res* 1996; 24:5064-6.

4. ***Hajkova P, El-Maarri O, Engemann S, Oswald J, Olek A, Walter J:*** DNA-methylation analysis by the bisulfite-assisted genomic sequencing method. *Methods Mol Biol.* 2002; 200:143-54.

5. ***Adorjan P, Distler J, Lipscher E, Model F, Muller J, Pelet C, Braun A, Florl AR, Gutig D, Grabs G, Howe A, Kursar M, Lesche R, Leu E, Lewin A, Maier S, Muller V, Otto T, Scholz C, Schulz WA, Seifert HH, Schwope I, Ziebarth H, Berlin K, Piepenbrock C, Olek A:*** Tumour class prediction and discovery by microarray-based DNA methylation analysis. *Nucleic Acids Res.* 2002; 30:e21.

6. ***Xiong Z, Laird PW:*** COBRA: a sensitive and quantitative DNA methylation assay. *Nucleic Acids Res.* 1997; 25:2532-4.

7. ***Eads CA, Danenberg KD, Kawakami K, Saltz LB, Blake C, Shibata D, Danenberg PV, Laird PW:*** MethyLight: a high-throughput assay to measure DNA methylation. *Nucleic Acids Res.* 2000; 28: E32.

8. ***Gonzalgo ML, Jones PA:*** Rapid quantitation of methylation differences at specific sites using methylation-sensitive single nucleotide primer extension (Ms-SNuPE). *Nucleic Acids Res.* 1997; 25:2529-31.

9. ***El-Maarri O, Herbiniaux U, Walter J, Oldenburg J:*** A rapid, quantitative, non-radioactive bisulfite-SNuPE- IP RP HPLC assay for methylation analysis at specific CpG sites. *Nucleic Acids Res.* 2002; 30:e25.

CHAPTER 9

ISOLATION AND SEQUENCING OF HETERODUPLEX FRAGMENTS FROM DHPLC ANALYSIS OF SOMATIC, LOW- FREQUENCY MUTATIONS

Jan Alsner

Contact:
Jan Alsner
Department of Experimental Clinical Oncology, Aarhus University Hospital,
Noerrebrogade 44, Bldg. 5, 8000 Aarhus C, Denmark
E-mail: jan@oncology.dk

0-9664027-7-4/03/$0.00+$.50 *From:* **Genetic Variance Detection~Nuts & Bolts of DHPLC in Genomics** (pp.163-174)
Edited by: Karl H. Hecker, Ph.D.

9.1 Introduction

Identification of somatic mutations in samples from solid tumors represent some additional challenges compared to the identification of heterozygous germline mutations. Biopsies from solid tumors often contain various amounts of normal cells and the tumor itself can be heterogeneous with respect to the mutation. Thus, the frequency of any one mutant allele is usually unknown. If the tumor is homogenous for a mutation in one allele of a gene and has lost the remaining wild-type copy of the gene, and the normal tissue component has been minimized through dissection of the sample, the frequency of the mutant allele can be very high. Otherwise, a heterogeneous tumor sample containing a large amount of normal tissue will have a very low frequency of the mutant allele. In either case, the heteroduplex fraction will be small and the DHPLC conditions must be optimized in order to reduce the amount of polymerase induced errors and, if possible, to sharpen the wild-type peak.

When a sample has been identified with a heteroduplex peak pattern, the mutation has to be characterized. If the frequency of the mutant allele is high, the sample can easily be sequenced directly. If the frequency is low, the heteroduplex peak(s) has/have to be isolated, re-amplified, and sequenced.

DNA fragments can be isolated with the WAVE® Nucleic Acid Fragment Analysis System. Previously described applications include subcloning of PCR products[1] and purification of template DNA for sequencing[2,3]. Here, a procedure for isolation, re-amplification and sequencing of low-frequency mutations is described. As examples two mutations in exon 8 of the p53 gene (*TP53*) are analyzed. This particular exon is selected based on its characteristic melting domain profile with a central 'high-melting' domain (Figure 9-1). The two mutations were detected using denaturing gradient gel electrophoresis (DGGE) and are selected based on their localization in different domains. Mutation 1 is a G>T located in the 5' 'low-melting' domain and is therefore expected to be easily identified by DHPLC. Mutation 2 is an A>T in the central 'high-melting' domain and is expected to be more difficult to recognize.

9.2 Results and Discussion

9.2.1 DHPLC Conditions

Genomic DNA was isolated from tumor samples as described previously[4]. Exon 8 of *TP53* was amplified using a nested PCR approach. In the first

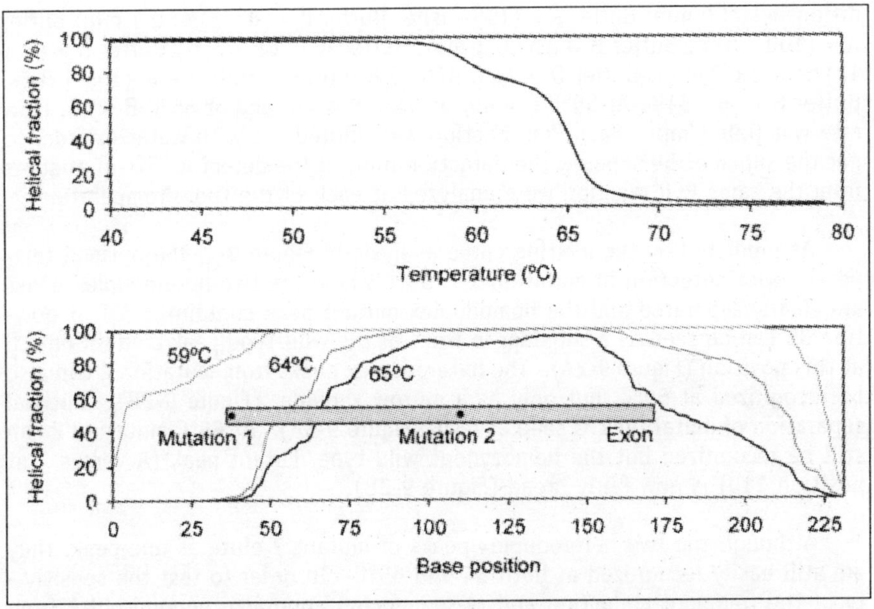

Figure 9-1: Melting profiles of TP53 exon 8 as predicted by the WAVEMAKER™ software version 4.1.16 (Transgenomic). The amplicon is 231 bp long and the exon part covers position 36 to 172. Mutation 1 is a G>T at position 38 and mutation 2 is an A>T at position 110.

round, approximately 10 ng of genomic DNA was amplified in a 25-µl reaction using 1.25 U of Optimase™ DNA polymerase (Transgenomic Inc., Omaha, NE) for 25 cycles (annealing temperature: 56°C). The primers used were 5'-AAATGGGCAGGTAGGACCTGAT (sense) and 5'-GTGCTAGGAAAGAGGCAAGGAA (antisense). The product of the latter reaction was used either as a template for the second round of PCR for the DHPLC analysis or for the re-amplification prior to the sequencing reaction as described below. The second round of PCR amplification was performed in a 25-µl reaction using 1.25 U of Optimase™ polymerase, but for 30 cycles (annealing temperature: 62°C). The primers used were 5'-ACTGCCTCTTGCTTCTCTTTTCC (sense) and 5'-AATCTGAGGCATAACTGCACCC (antisense). The anti-sense primer was labeled with 6-FAM (ABI) at the 5'-end.

DHPLC was performed on a WAVE® Nucleic Acid Fragment Analysis System from Transgenomic, which was equipped with a fluorescence detector (PMT voltage: high; response time: 0.1 sec) and controlled by the HSM software version 3.0. The buffers used were: buffer A: 0.1 M triethylammonium acetate (TEAA); buffer B: 0.1 M TEAA/25% acetonitrile (ACN); buffer D: 75% ACN. Gradient conditions were: 0 min: Buffer A = (105 - B)%, Buffer B = (B - 5)%; 0.1 min: Buffer A = (100 - B)%, Buffer B = B%; 6.1 min: Buffer A = (88 - B)%, Buffer B = (B + 12)%; 6.2-6.7 min: Buffer D = 75% ACN; 6.8-9.0 min: Buffer A = (105 - B)%, Buffer B = (B - 5)%. At 59°C B = 49, at 64°C B = 45, and at 65°C B = 44. Flow rate was 0.9ml/min. Each PCR reaction was diluted 1:1 with water in order to get the signal at 59°C below the detection limit of the detector. Six-µl aliquots from the same PCR reaction were analyzed at each of the three temperatures.

As predicted by the melting curve analysis in Figure 9-1, the optimal temperature for detection of mutation 1 is 59°C where the two heteroduplex peaks are clearly separated and the homoduplex mutant peak containing a T at position 38 can be seen as a shoulder in front of the wild-type peak containing a G at this position (Figure 9-2A). The heteroduplex peaks from mutation 2 can also be recognized at 59°C, but only as a narrow shoulder (Figure 9-2B). Optimal separation of mutation 2 is seen at 64°C (Figure 9-3A). At 65°C mutation 2 can still be recognized but the homozygous wild-type/mutant peak (A versus T at position 110) is now fairly broad (Figure 9-3B).

Although the two heteroduplex peaks of mutant 2 elute as one peak, they are still easily recognized at both 64 and 65°C. In order to test the sensitivity of the fragment collection and re-sequencing approach, genomic DNA from mutant 2 was diluted with increasing amounts of wild-type genomic DNA and analyzed together with the original mutant 2. One of these dilutions is included in Figures 9-3A and 9-3B.

9.2.2 Fragment Collection

Fragments were collected with a WAVE® Fragment Collector Model FCW 200 controlled by the Fragment Collector software version 1.0.0.1 (Transgenomic). For elution profiles like mutation 1 (Figure 9-2A) it is possible to use 'Peak' detection where the fraction is automatically collected around a given peak. For mutation 1, an approximately 100-µl fraction (corresponding to 0.11 min collection time) is collected around the initial heteroduplex peak. For profiles like mutation 2, where the fragment of interest is not a peak, an alternative strategy is to use a 'Whole Window' approach where approximately 50-µl fractions are collected every 0.1 min independent of the signal. In the example shown in Figure 9-3B, fractions were collected from 5.40 - 5.46, 5.50 - 5.56, and so on until 6.20 - 6.26 minutes.

Depending on the design of the fragment collector, a 'dead volume' may be present in the instrument. With the FCW 200 a small volume from the valve to the tip of the dispenser will be carried on to the next collection during multiple collections from a single sample and from the last collection of one sample

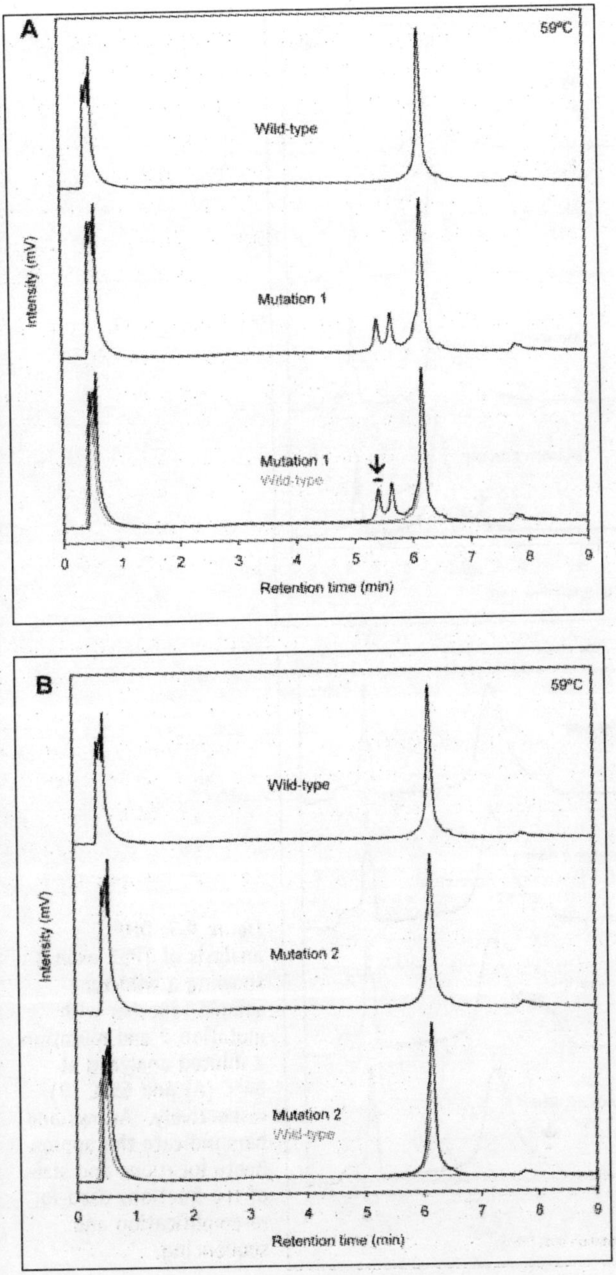

Figure 9-2: DHPLC analysis of TP53 exon 8 showing a wild-type sample together with mutation 1 (A) and mutation 2 (B) analyzed at 59°C. Arrow and bar indicate the approximate location and size of the fraction used for re-amplification and sequencing.

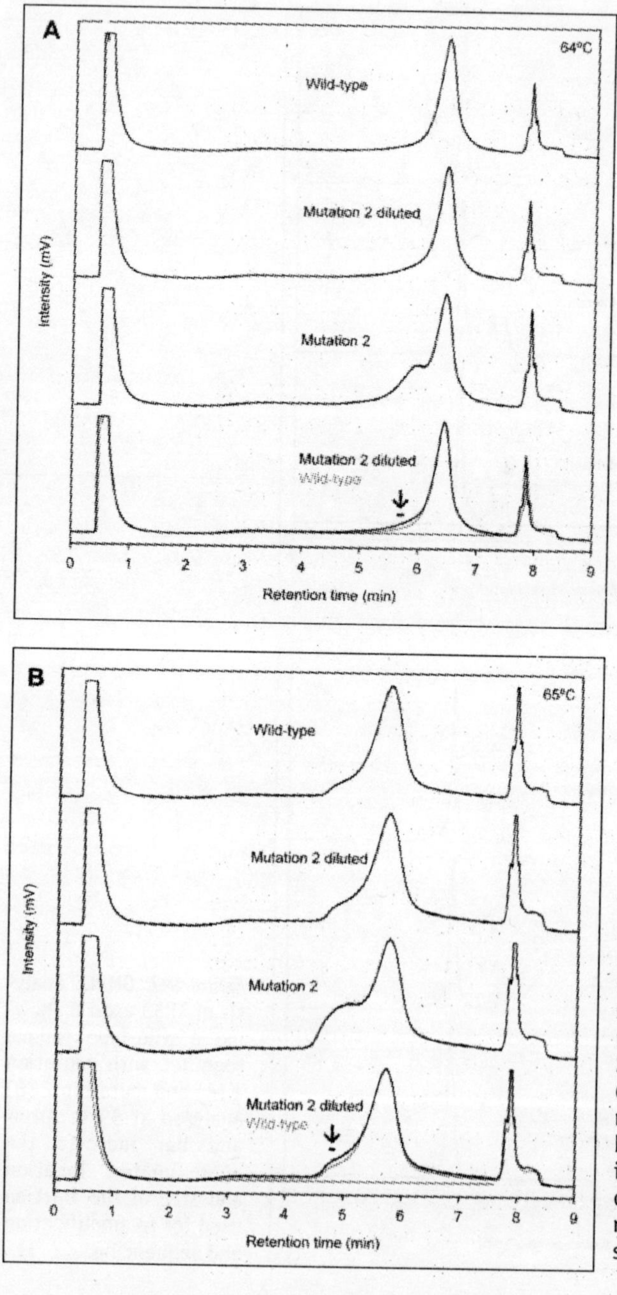

Figure 9-3: DHPLC analysis of TP53 exon 8 showing a wild-type sample together with mutation 2 and mutation 2 diluted analyzed at 64°C (A) and 65°C (B), respectively. Arrows and bars indicate the approximate locations and sizes of the fractions used for re-amplification and sequencing.

to the first collection of the next. To avoid any contamination from one sample to another, an initial fraction is collected from every sample before any attempts to isolate the heteroduplex peak(s). In the examples shown in Figures 9-2A, 9-3A, and 9-3B, this first fraction was collected from 2.00 - 2.11 minutes (approximately 100µl) where the signal is at baseline.

9.2.3 Re-sequencing of Isolated Fragments

The elution buffer in collected fractions does not affect any subsequent PCR reactions even when present at 20% by volume[2,5]. For the re-amplification of the collected heteroduplex fractions in Figures 9-2A, 9-3A, and 9-3B, 1 µl of the collected material was re-amplified in a 25-µl reaction using 1.25 U Optimase™ polymerase for 30 cycles. PCR products were purified from agarose gels using the QIAquick® Gel Extraction Kit (QIAGEN Inc.) prior to sequencing with ABI PRISM® BigDye™) Terminators v1.0 on an ABI PRISM® 310 Genetic Analyzer (Applied Biosystems).

In order to obtain uniform sequence data from various PCR products, all re-amplification reactions were performed with tailed PCR primers. Thus, for *TP53* exon 8 the sense primer is 5'-ctcctgttccgaccctgccACTGCCTCTTGCTTCTCTTTTCC and the anti-sense primer is 5'-cggaacaggagagcgcacgAATCTGAGGCATAACTGCACCC. The nucleotides in lower case represent the two universal sequencing tails/primers that are used for bi-directional sequencing. The exon specific part of each primer, shown in upper case, are identical to the primers used to generate the PCR products for the WAVE System analysis.

Initial attempts to use -21 M13 and M13 REV as universal sequencing primers did not provide satisfactory results (data not shown). Instead, the ABI PRISM® Primer Express™ software version 1.0 (Applied Biosystems) was used to identify a new set of primers that could also be used with the pGEM®-3Zf(+) plasmid which is provided as a positive control template with the BigDye(tm) Terminators kits.

Figure 9-4 shows sequencing results for mutation 1. The wild-type and mutation 1 'direct' sequences are derived from re-amplifications of the same initial first round of PCR that was used as a template for the PCR products analyzed on the WAVE System. Although it is present at a fairly low level, the G>T mutation in mutation 1 'direct' is clearly seen in both directions and would probably have been identified without the need for an isolation and re-amplification of the heteroduplex peak. However, the sequence of the mutation 1 'purified' illustrates how the intensity of the mutant signal can be significantly improved by the isolation/re-amplification approach.

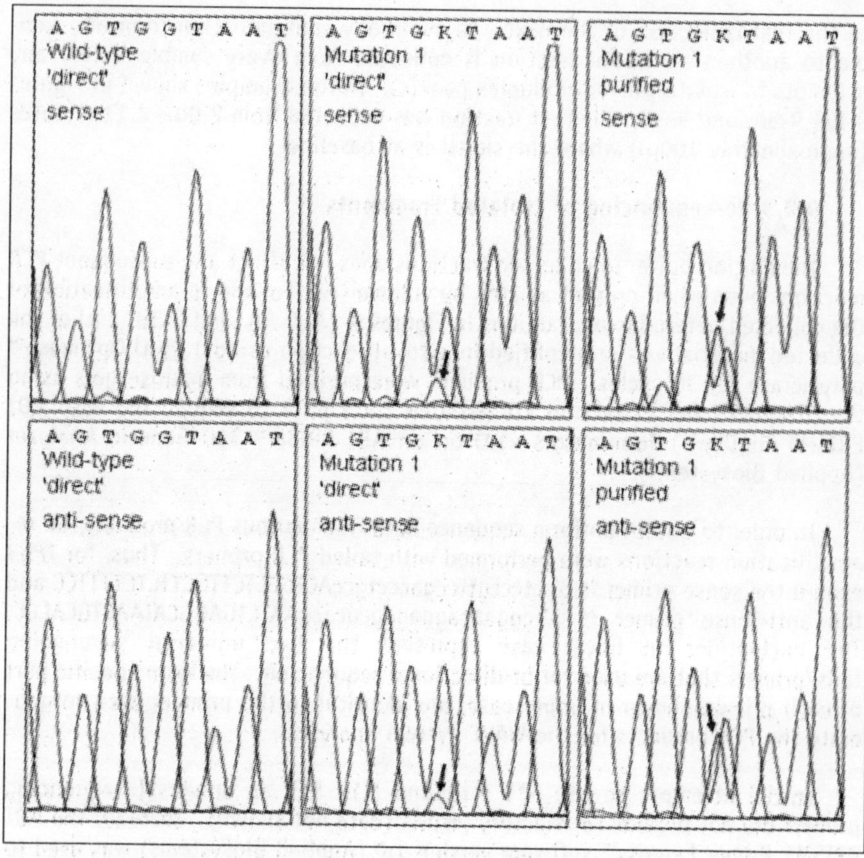

Figure 9-4: Sequencing results of a wild-type sample and mutation 1, either directly or following purification and re-amplification of the initial heteroduplex peak.

Figure 9-5 shows sequencing results for mutation 2. In this case, the very faint signal in the mutant 2 diluted 'direct' sequencing result would not have been recognized as more than random noise. However, even though the fraction collected at 64 °C includes a significant proportion of wild-type molecules (the signal of the true wild-type is elevated from the base-line, see Figure 9-3A), the mutant fraction is enriched to a point where it can be identified in the sequence analysis. At 65°C (Figure 9-3B), the proportion of wild-type molecules is even higher and the mutant signal in the sequence is only slightly improved.

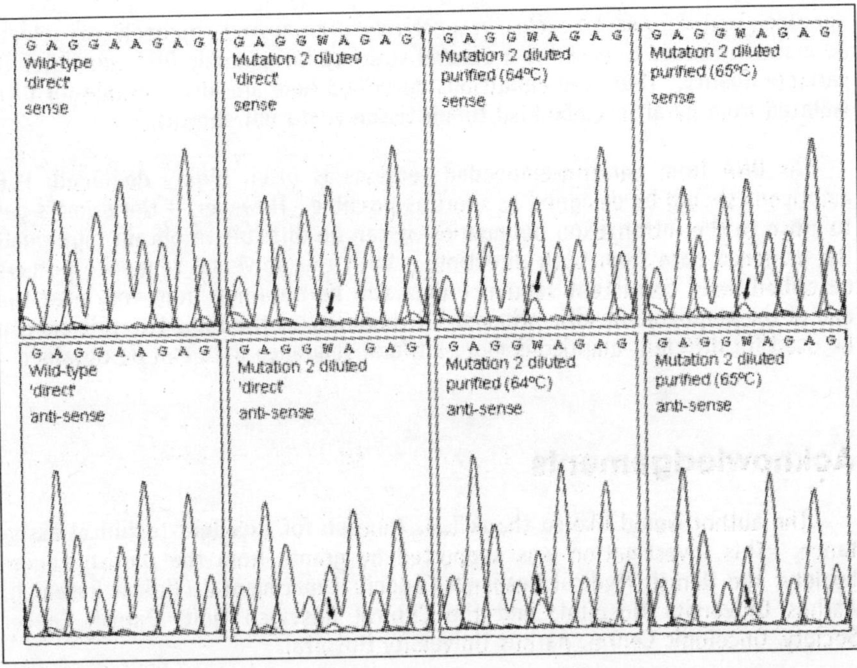

Figure 9-5: Sequencing results of a wild-type sample and mutation 2 diluted, either directly or following purification and re-amplification of the heteroduplex shoulder identified at either 64 or 65°C.

9.3 Conclusion

DHPLC is a sensitive method for the detection of the presence of low-frequency mutations that may not be reliably recognized by direct sequencing. Combined with a fragment collector, it is also possible to perform automated isolation of small heteroduplex peaks allowing for the characterization of these mutations by re-amplification and sequencing.

The sensitivity of DHPLC for detection of low-frequency mutations and the quality of the sequence data obtained from small heteroduplex peaks depend on the quality of the PCR product and the melting profile. Any errors introduced by the polymerase during the PCR reaction will accumulate as a broad peak in front of the wild-type peak and lower the ratio of the true mutation in the sequence analysis. As demonstrated here, Optimase™ polymerase can provide PCR products virtually free of polymerase-induced errors even after a nested approach with 25 and 30 cycles, respectively. The nested approach was selected mainly in order to describe a general strategy for genomic DNA templates of variable quality. Thus, the conditions described here are also suitable for DNA isolated from paraffin-embedded tumor tissue (data not shown).

As DNA from paraffin-embedded sections is often highly degraded, PCR amplicons should be designed as short as possible. However, if the primers get to close to the intron/exon boundaries, it can be difficult to obtain high quality sequence data from both directions. With the universal sequence primers described here, consistent sequence data can be obtained from very near the end of the amplicon-specific primers. Furthermore, as the same two primers can be used for different amplicons more uniform sequence data can be obtained.

Acknowledgements

The author would like to thank I.M. Thuesen for excellent technical assistance. This investigation was supported by grants from the Danish Cancer Society; the Danish Medical Research Council; Experimental Clinical Research, Aarhus University Hospital; and the Clinical Research Unit, Danish Cancer Society, Oncologic Center, Aarhus University Hospital.

References

1. ***Shaw-Bruha CM, Lamb KA:*** Ion pair-reversed phase HPLC approach facilitates subcloning of PCR products and screening of recombinant colonies. *Biotechniques* 2000;28:794-797.

2. ***Shaw-Bruha CM, Kuklin A, Taylor P:*** PCR amplification and automated sequencing of DNA fragments isolated with the WAVE® Nucleic Acid Fragment Analysis System. Transgenomic Inc., *Application Note 104*, http://www.transgenomic.com

3. ***Hecker KH, Kobayashi K***: Sequencing of DNA fragments isolated with the WAVE(r) Nucleic Acid Fragment Analysis System. Transgenomic Inc., *Application Note 115*, http://www.transgenomic.com

4. ***Alsner J, Yilmaz M, Guldberg P, Hansen LL, Overgaard J:*** Heterogeneity in the clinical phenotype of TP53 mutations in breast cancer patients. *Clin Cancer Res* 2000;6:3923-3931.

5. ***Kuklin A, Munson K, Taylor P, Gjerde D:*** Isolation and analysis of amplified cDNA fragments during detection of unknown polymorphisms with temperature modulated heteroduplex chromatography. *Mol Biotechnol* 1999;11:257-261.

CHAPTER 10

ANALYSIS OF PROMOTER METHYLATION USING DENATURING HIGH-PERFORMANCE LIQUID CHROMATOGRAPHY (DHPLC)

Beate Betz and Dieter Niederacher

Department of Gynecology and Obstetrics, University Hospital Düsseldorf, Germany

Contact:
Beate Betz, Dr. rer. nat.
Molekulargenetisches Labor, Frauenklinik der Heinrich-Heine-Universität,
Gebäude 23.12.04 , Moorenstraße 5, 40225 Düsseldorf, Germany
E-mail: betzb@uni-duesseldorf.de

0-9664027-7-4/03/$0.00+$.50 *From:* **Genetic Variance Detection~Nuts & Bolts of DHPLC in Genomics** (pp.175-186)

Edited by: Karl H. Hecker, Ph.D.

10. 1 Introduction

Sequencing and analysis of the human genome revealed the presence of only 30,000 to 40,000 protein-coding genes[1], a stark contrast to previous estimates of about 100,000 genes. While a large number of putative tumor suppressor genes (TSG), which are down regulated in cancer development, could be identified by array technology, it remains largely problematic to demonstrate possible inactivation mechanisms for putative TSG identified. A widely known and accepted mechanism of tumor suppressor gene inactivation in mammalian cell cancer is the methylation of CpG islands, particularly in promoter regions, leading to the formation of 5-methyl cytosine. Two techniques are commonly employed for the analysis of promoter methylation pattern. Methylated and unmethylated genomic DNA can be distinguished by either the use of methylation-sensitive restriction enzymes or by conversion of genomic DNA with sodium bisulfite. Sodium bisulfite converts unmethylated cytosine (C) to uracil (U). Methylated C is not affected by the chemical treatment with bisulfite. The effects of bisulfite conversion of genomic DNA can be evaluated by several more or less time consuming or sensitive assays. Such assays include subcloning and subsequent sequencing, restriction analysis, as well as other techniques (see http://www3.mdanderson.org/leukemia/methylation/ and references cited therein)

Here we describe the use of Denaturing High Performance Liquid Chromatography (DHPLC) as an alternative screening technology for promoter methylation. The technology was first described by Baumer *et al.*[2] in the context of investigating the methylation status of imprinted genes. DHPLC combines several advantages. It is a highly sensitive high-throughput technique. The technology, which is applicable to fragments ranging in size from approximately 200 to 400 bp, not only facilitates the simultaneous analysis of multiple CpG sites, it also permits differentiation between different methylation patterns. Thus, it becomes possible to distinguish between different methylation pattern such as single-site CpG methylation and complete methylation of all CpG sites within a given fragment. Consequently, verification and characterization of the methylation status on the base level has to be performed only on DHPLC-positive samples by PCR and/or subcloning followed by sequencing.

The fundamental principle of the DHPLC technique for mutation detection and methylation analysis is sequence-dependent denaturation of DNA fragments in the presence and absence of mismatches.[3] In the case of methylation analysis genomic DNA is initially treated with sodium bisulfite, which is followed by PCR amplification, and final analysis of PCR products under partially denaturing conditions. In the case where a single CpG-site is methylated, causing protection of that site from the effects of bisulfite treatment, the resulting

PCR product will contain only one cytosine. All unmethylated cytosines will have been converted to thymidine in the final PCR product. After denaturation/ renaturation differences in the methylation status of a given site will result in heteroduplex formation between respective PCR products. In the case where differences in methylation pattern are limited to a single site, the detection of these differences is equivalent to the detection of point mutations. Complete methylation of all CpG-sites within an amplified region results in an increased GC content of the amplified DNA fragment after sodium bisulfite treatment compared to the PCR product derived from unmethylated genomic DNA. Increased GC content results in increased stability and melting temperature of the PCR fragment analyzed, which in turn translates into increased retention times of respective PCR products during DHPLC analysis under partially denaturing conditions. Analysis of genomic DNA representing different methylation pattern will result in PCR fragments of differing sequences after bisulfite treatment, each of which may or may not have a distinctive melting temperature and retention time. Furthermore, a more or less complex mixture of heteroduplexes is expected to form. This will result in complex DHPLC elution profiles reflecting the complexity of methylation pattern in the PCR-amplified region of interest.

10.2 Methylation Analysis by DHPLC

Implementation and testing of DHPLC as a prescreening method for methylation analysis was performed on a 392-bp promoter fragment containing 35 CpG sites. In our opinion and based on our experience the size of a fragment analyzed should be between 200-400 bp, because fragments of less than 200 bp may not give satisfactory results, as the fragments melt over a very narrow temperature range. Analysis of fragments larger than 400 bp may reduce the sensitivity of the method due to the multitude of heteroduplexes that may be formed. The latter will be discussed in more detail later. Moreover, it is difficult to amplify sequences greater than 600 bp in length from bisulfite-treated DNA due to bisulfite-induced strand breakages.

Primers and PCR conditions were optimized to amplify a PCR product from the promoter region of sodium bisulfite converted, methylated and unmethylated DNA. No PCR product needs to be amplified from untreated genomic DNA. Therefore, CpG dinucleotides in the primer sequence have to be avoided. However, several bases that are originally present as cytosines at the 3′-end of the primers will help ensure that unconverted sequences are not amplified. Amplification of the promoter region by PCR benefits from the use of a low-fidelity polymerase. High-fidelity polymerases, usually preferred for DHPLC analysis, generally do not tolerate the presence of uracil in template DNA. Thus, high-fidelity polymerases are not suitable for the direct amplification from bisulfite-treated DNA, at least during the initial stages of the PCR. On the other hand, a large number of PCR cycles or two consecutive rounds of PCR amplification are often required in order to generate sufficient PCR product for analysis due to low PCR efficiency. In order to nevertheless achieve a PCR product with a low overall rate of misincorporation, addition of a proofreading polymerase may be beneficial after the initial cycles of a PCR.[4]

Several DNA samples with different, well-characterized methylation pattern for the chosen promoter were investigated. Two tumor cell lines that had previously been characterized as fully methylated or unmethylated in the promoter region analyzed and DNA derived from a tumor showing methylation of twelve of the 3′-terminal CpG sites were available for testing. Additionally, positive controls with only a few, well defined methylated CpG sites were generated using specific methylases. While HhaI methylase modifies the internal cytosine residue in the sequence 5′-GCGC-3′, HpaII methylase modifies the internal cytosine residue in the sequence 5′-CCGG-3′. Unmethylated genomic DNA was treated with HpaII and/or HhaI. Use of the individual methylases results in the methylation of three sites each in the promoter region of interest. When used in combination, a total of six sites are methylated in the promoter region of interest. Complete methylation of the promoter fragment of interest can be achieved by treatment of unmethylated genomic DNA with SssI methylase, which modifies cytosine residues in the sequence 5′-CG-3′. Mixing of methylase-treated DNA with wild type DNA is not necessary to achieve heteroduplex formation, because a complete conversion of all putative methylation sites is not probable. DNA samples derived from the blood of young human subjects can be used as controls representing completely unmethylated DNA. Genomic DNA from young human subjects should be used in order to avoid the effects of age-related methylation. In this way it is possible to generate positive controls for every promoter fragment of any putative TSG known so far, which enables one to optimize specifically conditions for DHPLC analysis.

10.3 Choosing DHPLC Analysis Conditions

Several issues need to be considered when determining optimal running conditions for DHPLC analysis on the WAVE® Nucleic Acid Fragment Analysis System (Transgenomic, Omaha, NE). Optimal gradient conditions and analysis temperatures are determined using WAVEMAKER™ software (Transgenomic, Omaha, NE). The software is used to predict melting curves for PCR products resulting from fully methylated, i.e.: all cytosines are conserved in the sequence of interest, and completely unmethylated DNA, i.e.: all cytosines in the sequence of interest are converted to thymidine, after bisulfite treatment. It is apparent that the melting temperature of the PCR product resulting from fully methylated genomic DNA has a higher stability due to its increased GC content relative to unmethylated DNA. While the PCR product resulting from unmethylated genomic DNA after bisulfite treatment is partly denatured at 61°C, the corresponding fragment from methylated genomic DNA still remains largely double stranded. At 63°C the latter fragment is only partly denatured while the PCR amplicon resulting from unmethylated genomic DNA after bisulfite treatment is predominantly in its single-stranded form (Figure 10-1).

Thus, the optimal temperature for detecting a single methylated CpG site in the fragment analyzed here is predicted to be 61°C. Considering that methylation of more than one CpG site is generally expected, the resulting number of anticipated base changes will result in destabilization of the DNA helix and a concomitant lowering of the melting temperature of the fragment analyzed. Thus, the optimal temperature for DHPLC analysis should be in a temperature range below the WAVEMAKER software predicted temperature for the fully unmethylated region of interest. At this temperature a shift in retention time between the PCR product derived from methylated and unmethylated fragments can be observed. However, the differences in retention times are more obvious

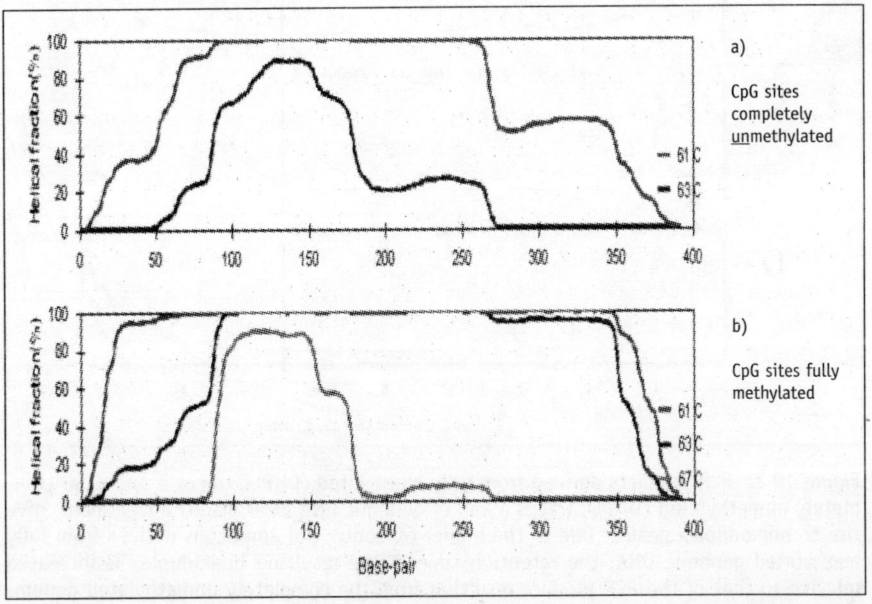

Figure 10-1: WAVEMAKER software predicted melting curves for a 392-bp promoter fragment. The melting behavior for the completely unmethylated (**a**) and the fully methylated fragment (**b**) reveal increased stability of the double-stranded DNA due to higher GC content in the fragment derived from fully methylated genomic DNA after bisulfite treatment.

at higher temperatures, i.e. the melting temperature predicted for the PCR product resulting from the fully methylated region of interest, which is at approximately 63°C in the fragment analyzed here. The differences in retention times can be correlated directly to the different degrees of denaturation between fragments. This is a direct consequence of the higher affinity of the DNASep™ Cartridge (Transgenomic, Omaha, NE) for double-stranded DNA as compared to single-stranded DNA (Figure 10-2).

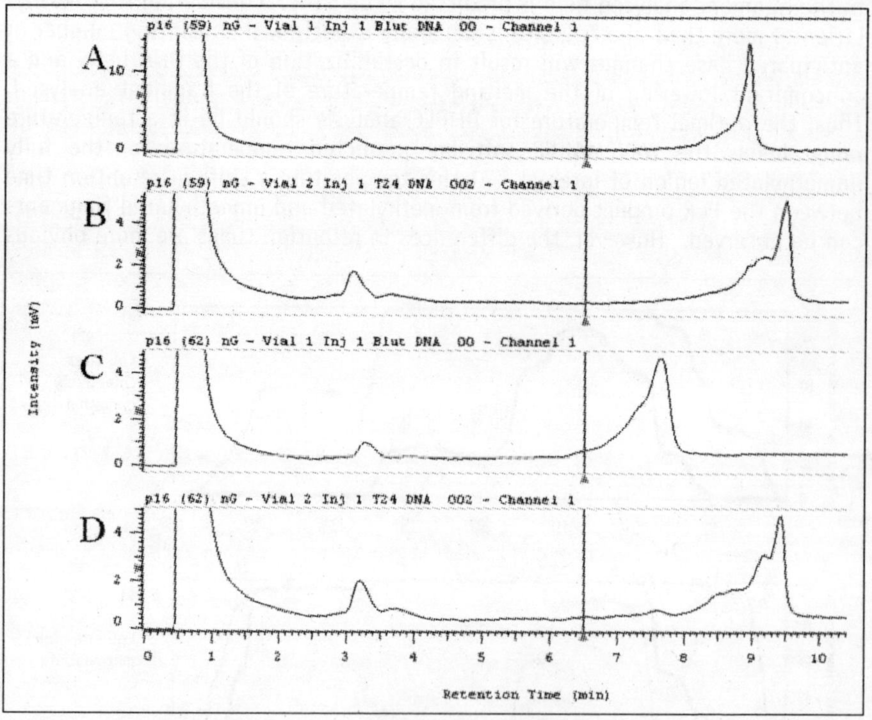

Figure 10-2: PCR products derived from fully methylated (DHPLC traces b and d) or completely unmethylated (DHPLC traces a and c) genomic DNA after bisulfite treatment, give rise to homoduplex peaks. Due to the higher GC content of amplicons derived from fully methylated genomic DNA, the retention time of the resulting homoduplex is increased relative to that of the PCR product resulting from the completely unmethylated genomic DNA. The observed shift in retention time is more pronounced at higher temperatures, i.e. the predicted melting temperature of the GC-rich fragment (DHPLC traces c and d). The latter observation can be explained by the fact that at this higher temperature the GC-rich fragment still retains significant double-stranded character, while the corresponding AT-rich fragment is largely melted.

When determining gradient conditions for methylation analysis by DHPLC with the aid of the WAVEMAKER software, one needs to consider that multiple methylation patterns may be present. This is particularly the case when analyzing genomic DNA from tumors, which may contain different methylation patterns in and by themselves, but may also contain genomic DNA from surrounding healthy tissues. Heterogeneity in the methylation pattern of the genomic DNA analyzed will lead to extensive heteroduplex formation. Contrary to the hypothesis proposed by Deng *et al.*[5], which assumes that the presence of multiple sequence differences inhibit heteroduplex formation, it is our opinion that this is not the case. We conclude from our observations and experience that heteroduplex formation is efficient even if multiple sequence variants are present. The ratio of homo- and heteroduplex formation is, however, dependent on the relative amounts of sequence variants present. If a particular sequence or sequences is/are dominant in a given mixture of sequence variants, this will result in prominent homoduplex peaks. Sequence variants present at low abundance will in this case form heteroduplexes efficiently, albeit at low abundance and with a wide spread of retention times. Increasing the elution gradient to 3 - 4% per minute of buffer B, which contains 25% acetonitrile, instead of the 2% per minute slope of the standard gradient recommended by WAVEMAKER software, has the effect of causing the elution of heteroduplexes formed in a shortened time interval. This leads to a decrease in resolution between heteroduplexes, but increased peak intensity of the mixture of heteroduplexes present. Using an elution gradient that spans a wider range of acetonitrile concentration than that recommended by WAVEMAKER software is beneficial in that hetero- and homoduplex signals are guarantied to elute within the analysis portion of the elution gradient (see ***Table 10-1***). Analysis of DHPLC results can further be improved by increasing the injection volume from usually 5μl, at a concentration of approximately 20ng of PCR product/μl, to 10-15μl. Quantity and quality of PCR products must be confirmed prior to methylation analysis by DPHLC. Quantification and quality determination of PCR products can be performed on the WAVE Nucleic Acid Fragment Analysis System quantitatively and by gel electrophoresis semi-quantitatively.

10. 4 Analysis of DNA Samples with Different Methylation Patterns

PCR products derived from fully methylated or completely unmethylated genomic DNA after bisulfite treatment result in the formation of a single homoduplex for each sample. As shown above, the retention times observed differ due to differences in melting temperatures as a function of GC content of the fragment analyzed. The PCR fragment with the higher GC content, i.e.: derived from fully methylated genomic DNA, exhibits an increased retention time over the PCR amplicon derived from completely unmethylated genomic DNA. The presence of different methylation patterns in the genomic DNA analyzed leads to the formation of PCR products of differing sequences. Hence, heteroduplexes are formed after denaturation and reannealing of the PCR products. The latter is demonstrated in a model system using genomic DNA treated with different combinations of the methylases HpaII and HhaI followed by sodium bisulfite conversion. Three or six CpG sites, respectively, are methylated by the individual methylases or the combined action of both. While peak shapes/heteroduplex pattern observed are very similar for genomic DNA treated with the individual methylases HpaII and HhaI (Figure 10-3b and c), the heteroduplex pattern of the PCR products generated from genomic DNA treated with both enzymes is distinctly different (Figure 10-3d) from the patterns shown in Figure 3, panel b and c. In the case of the heteroduplex pattern shown in Figure 3 e, which is derived from tumor DNA, the heteroduplex pattern generated appears to be more complex and spans a wider retention-time window than in the examples shown above. From this it is concluded that at least 12 CpG sites are methylated in the tumor's genomic DNA. From the examples shown so far, it is apparent that based on the elution profiles and peak patterns it is not only possible to distinguish between fully methylated and completely <u>un</u>methylated genomic DNA, but it is also possible to qualitatively determine the extend and diversity of methylation patterns present in the genomic DNA sample analyzed. Determination of the various heteroduplex patterns require a clear wild type signal obtained by complete conversion of unmethylated cytosines to uracil by bisulfite. The conditions for optimal bisulfite conversion of genomic DNA are discussed in detail by Warnecke *et al.*[4] It is essential that formation of double-stranded DNA during bisulfite conversion be avoided and that the DNA analyzed is of high quality.

10. 5 Analysis of Heterogeneous Tumor Material

In DNA derived from tumor material the methylation patterns can be extremely variable due to the heterogeneity of the tumor material. As mentioned above, the mixture of cells with complete CpG-site methylation, wild type cells without CpG-site methylation, and cells with methylation of different degrees result in formation of the two homoduplexes with distinct retention times as well as a mixture of heteroduplex peaks (Figure 10-4). Dependent on the portion of the cells with variable degrees of methylation the different peak signals can be more or less intensive, making it necessary to inject an increased amount of PCR product.

10.6 Conclusion

DHPLC is a suitable prescreening technology for the analysis of promoter methylation. Due to its high-throughput capability, DHPLC is a quick and cost effective alternative to conventional methods. This is particularly the case

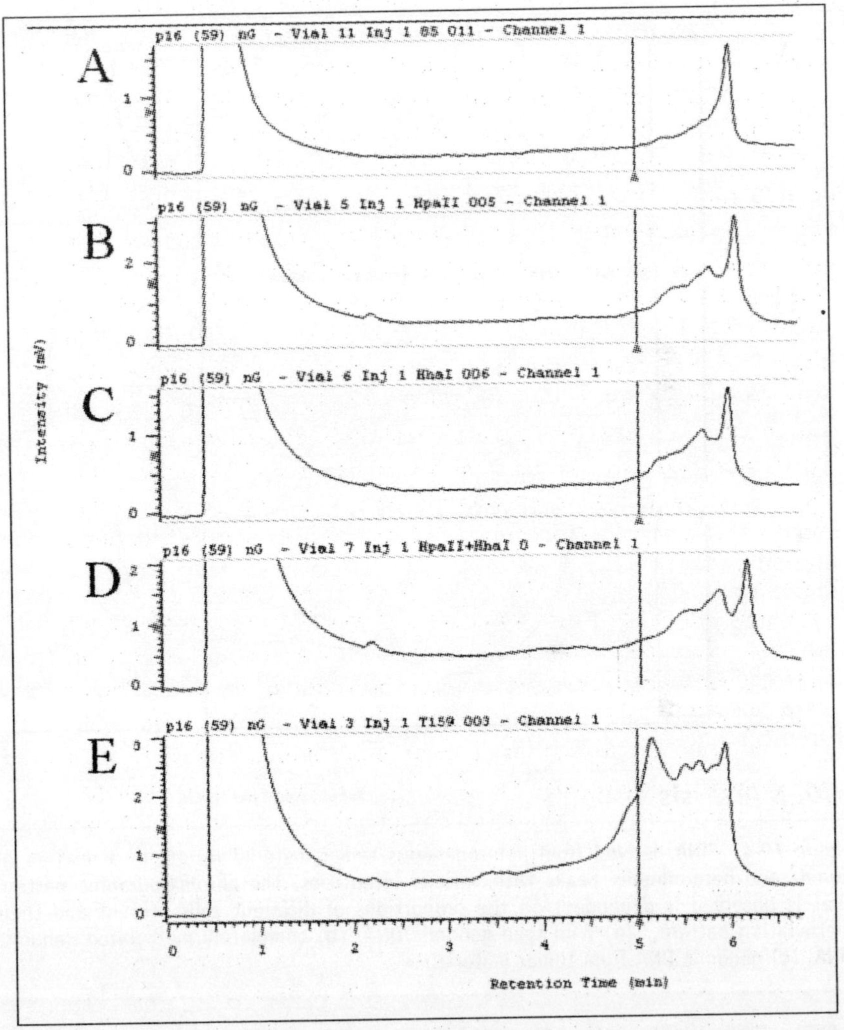

Figure 10-3: Different degrees of CpG-site methylation generated by using specific methylases result in various patterns of heteroduplex formation. The extent of heteroduplex formation increases with increasing numbers of methylated CpG sites in the amplified PCR fragment. The PCR product from unmethylated wild type DNA is shown in panel (**a**). DNA treated with either HpaII or HhaI alone are shown in panel (**b**) and (**c**), respectively. The heteroduplex pattern obtained after treatment of unmethylated genomic DNA with a combination of the methylases HpaII and HhaI is shown in panel (**d**). The heteroduplex pattern of DNA derived from a tumor cell line having the last 12 CpG sites methylated is shown in panel (**e**).

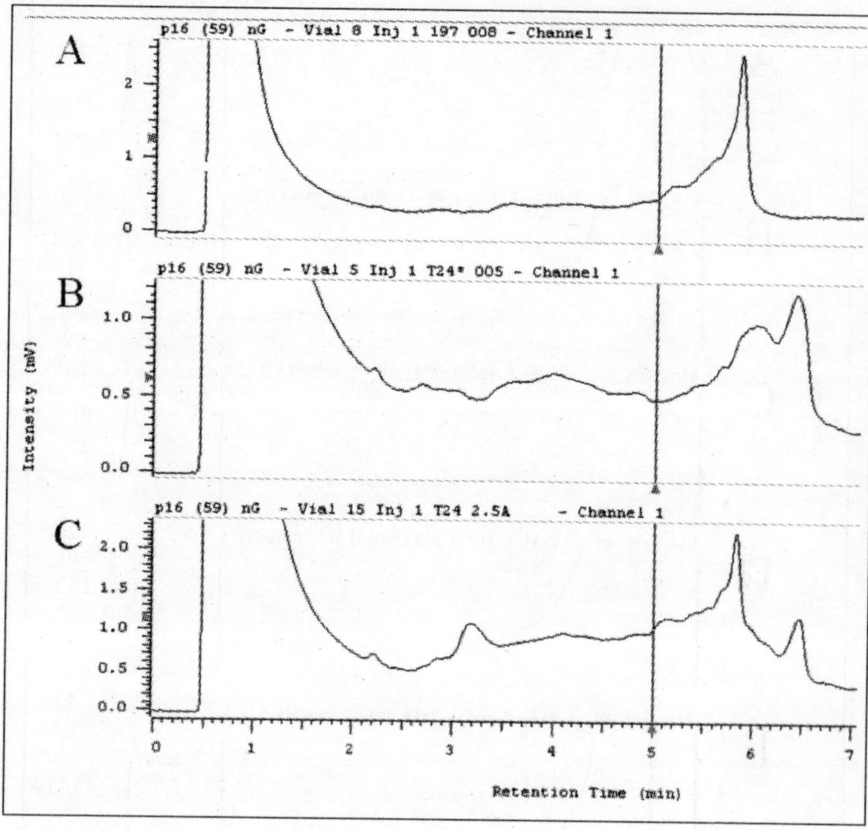

Figure 10-4: DNA derived from heterogeneous tumor material results in a mixture of homo- and heteroduplex peaks with various intensities. The chromatographic pattern that is observed is dependent on the proportions of different cells present and their methylation pattern. (**a**) Wild type genomic DNA, (**b**) completely methylated genomic DNA, (**c**) genomic DNA from tumor material.

when a great number of so far unknown putative tumor suppressor genes need to be screened for their methylation status in order to determine promoter methylation as a possible inactivation mechanism. Unlike restriction analysis, which is limited to the restriction enzyme recognition site, the DHPLC technology enables the determination of the methylation status at multiple CpG sites simultaneously within an amplified PCR fragment. Compared to subcloning and sequencing, DHPLC is significantly less laboriously and quicker. The DHPLC technology provides information about the general extend of methylation, i.e.: whether a fragment is fully methylated, completely unmethylated, or partially methylated. Sequencing of PCR products of DHPLC-positive samples, either directly or after cloning, following bisulfite treatment of genomic DNA is still required in order to determine the methylation pattern with resolution on the level of individual bases.

The high sensitivity of DHPLC enables the characterization of single CpG-site methylation as well as complete CpG-site methylation within a fragment analyzed. Even the degree of CpG-site methylation can be estimated up to a certain point by the extent of heteroduplex formation. Using specific methylases, positive controls for completely methylated (SssI) or partly methylated genomic DNA (HpaII or HhaI) can be generated. Thus, before analyzing unknown promoter fragments optimal experimental conditions for the analysis by DHPLC, like gradient conditions and analysis temperatures, can easily be established.

Table 10-1: DHPLC gradient conditions and analysis temperatures

	Software Predicted Gradient (61°C)		Optimized Gradient I (59°C)		Optimized Gradient II (62°C)	
Step	**Time**	**%B**	**Time**	**%B**	**Time**	**%B**
Loading	0.0	49	0.0	47	0.0	44
Start Gradient	0.5	54	0.5	52	0.5	49
Stop Gradient	5.0	63	5.5	67	6.5	67
Start Clean	5.1	100	5.6	100	6.6	100
Stop Clean	5.6	100	6.1	100	7.1	100
Start Equilibrate	5.7	49	6.2	47	7.2	44
Stop Equilibrate	8.2	49	8.2	47	9.2	44

Table 10-1: Comparison between gradient conditions and analysis temperatures predicted with the WAVEMAKER software and empirically optimized conditions. Buffer B, 25% acetonitrile (ACN), 0.1 M triethylammonium acetate (TEAA) increases at a rate of 3%/min in the empirically optimized gradient instead of the 2 %/min slope recommended by WAVEMAKER software. The analysis temperature for methylation analysis was lowered in comparison to the software recommended analysis temperature for mutation detection. The lower analysis temperature is rationalized by the possible presence of multiple sequence variations in the PCR product analyzed due to the possible presence of multiple methylation patterns.

References

1. International Human Genome Sequencing Consortium, Initial sequencing and analysis of the human genome. *Nature* 405, 860-921 (2001)

2. ***Baumer A, Wiedemann U, Hergersberg M and Schinzel A:*** A novel MSP/DHPLC method for the investigation of the methylation status of imprinted genes enables the molecular detection of low cell mosaicisms. *Human Mutation* 17, 423-430 (2001)

3. ***Huber C. H:*** Micropellicular stationary phases for high-performance liquid chromatography of double stranded DNA. *Journal of Chromatography* A 806, 3-30 (1998)

4. ***Warnecke PM, Stirzaker C, Song J, Grunau C, Melki JR and Clark SJ:*** (2002). Identification and resolution of artifacts in bisulfite sequencing. *Methods* 27:101-107.

5. ***Deng D, Deng G, Smith M F, Zhou J, Xin H, Powell SM and Lu Y:*** Simultaneous detection of CpG methylation and single nucleotide polymorphism by denaturing high performance liquid chromatography. *Nucleic Acids Research* 30, e13 (2002)

INDEX

Acetonitrile, 17, 37, 60, 141
Alu I, 97
AmpliTaq, 38
Apolipoprotein B (Apo B), 118

B-cell precursor, 139
Bisulfite, 152, 176
Bone marrow (BM), 139
Bovine serum albumin, 21
BRCA1, 54-65
BRCA2, 54-65
Breast Cancer Information Core, 55
Burt-Hogg-Dube syndrome, 38

CHR, 96-101
Clonal antigen receptor rearrangements, 138
Coronary heart disease, 118
CpG, 152, 176
CSGE, 83
Cystic fibrosis, 118

DGGE, 57, 81,
Dideoxy fingerprinting (DDF), 126
DNASep cartridge, 17, 26, 180

Familial Hypercholestrolaemia (FH), 118
Fragment collection, 140

German Consortium for Hereditary Breast and Ovarian Cancer, 51

HaeIII, 98
Heparin, 27
Heterogeneous tumor sample, 164
HhaI, 178-183
Homoduplexes (DNA), 20
Hot Start PCR, 37
HpaII, 178-183

K562 cell line, 95

Ligation During Amplification (LDA), 122
Low density lipoprotein receptor (LDLR), 118
Lymphoblastic leukemia (ALL), 138

Methylation Analysis, 152, 176-178
Minimal residual disease (MRD), 138
Mitochondria, 94-116
MITOMAP, 95

Navigator software, 17
NIST, 95
NP40, 26

Optimase Polymerase, 27, 38-48, 165, 171

p53, 164
Paraffin-embedded tumor tissue, 172
Pfu-Turbo, 38
Phenol, 26
Polyethylene glycol (PEG), 27
Polymerase fidelity, 38-39
Polyvalent metal ions, 26
Protein Truncation Test (PTT), 57

QIAGEN Taq, 41

Random Mutagenesis-PCR, 38
RFLP, 96

SDS, 26
Single nucleotide polymorphism (SNP), 16, 96
SNuPE techniques, 153
Solid tumors, 164
Somatic mutations, 164
SSCP, 57, 81, 83, 94, 119, 121
SssI methylase, 178
Standard Reference Material, 106

TaqI methylases, 178
T-Cell Receptor, 138
TEAA - triethyl ammonium acetate, 17, 37, 96
TGGE, 58, 94
Thermo Sequenase DNA Polymerase, 153
TMHA, 17
TOPO clone, 45
Triton X-100, 26
Tuberous sclerosis (TSC), 80-90
Tumor suppressor gene (TSG), 176
Tween 20, 26
Two- dimensional gene scanning (TDGS), 57

RNA INTERFERENCE
Nuts & Bolts

David R. Engelke, Ph.D.
Editor

ISBN: 0-9664027-8-2

This book offers protocols and advice for the use of RNA interference for knocking down expression of target genes in a number of different eukaryotes. Varied applications in mammalian systems are explored in depth by authors currently working at the leading edge of RNA interference methodologies. Chapters will include:

- RNAi in plants
- RNAi in *Drosophila*
- RNAi in *C. elegans*
- RNAi in mammalian systems
- Design, synthesis and preparation of synthetic RNAi
- Expression of RNAi from recombinant DNA and viral vectors
- High-throughput RNAi applications
- Emerging alternative RNAi methods
- RNAi in mice

For more info visit: www.dnapress.net

DNA Press™